# Engineering Materials in Extreme Environments

# Engineering Materials in Extreme Environments

Guest Editors

**Yi Gong**
**Qi Tong**

MDPI

Basel • Beijing • Wuhan • Barcelona • Belgrade • Novi Sad • Cluj • Manchester

*Guest Editors*
Yi Gong
Department of Materials Science
Fudan University
Shanghai
China

Qi Tong
Department of Aeronautics and Astronautics
Fudan University
Shanghai
China

*Editorial Office*
MDPI AG
Grosspeteranlage 5
4052 Basel, Switzerland

This is a reprint of the Special Issue, published open access by the journal *Materials* (ISSN 1996-1944), freely accessible at: www.mdpi.com/journal/materials/special_issues/Materials_Extreme.

For citation purposes, cite each article independently as indicated on the article page online and as indicated below:

| Lastname, A.A.; Lastname, B.B. Article Title. *Journal Name* **Year**, *Volume Number*, Page Range. |
|---|

**ISBN 978-3-7258-3916-2 (Hbk)**
**ISBN 978-3-7258-3915-5 (PDF)**
**https://doi.org/10.3390/books978-3-7258-3915-5**

# Contents

# Preface

The combination of diverse engineering materials, such as metals, polymers, ceramics, and their composites, with extreme service environments, including high temperature, high pressure, mechanical loads, chemicals, radiation, and their interactions, poses significant challenges to the reliability, safety, longevity, and economy of industrial equipment, which can lead to material degradation and failure and, ultimately, catastrophic consequences if not properly addressed. Therefore, a comprehensive understanding of engineering material performance under such extreme conditions is crucial for the successful operation of industrial equipment.

This Special Issue, entitled "Engineering Materials in Extreme Environments", aims to gather cutting-edge research and provide a comprehensive overview of the structures, properties, processing, and performance of engineering materials in extreme environments. By bringing together experts from various fields, it seeks to advance the understanding of material behaviors under extreme conditions and promote the development of new materials and technologies that can withstand these challenges.

The 11 peer-reviewed papers published in this Special Issue cover a wide range of topics, from fundamental research on material properties to practical studies on component performance in specific industries of civil engineering, high-speed trains, new energy, nuclear power, oil and gas, petrochemicals, tunnel construction, etc. These papers include experimental investigations, numerical simulations, and failure analysis case studies of the engineering equipment/materials of carbon fiber composites, cladded stainless steel plates, controllable bonded strands, electrical cables, enhanced tubes, extraction columns, functionally graded materials, lithium-ion batteries, nuclear fuel assemblies, shield machine cutterheads, and track circuit reader antenna baffles, which demonstrate the practical applications of these materials and provide valuable insights into material science and engineering in extreme environments.

We would like to express our sincere gratitude to all the authors who contributed their valuable research to this Special Issue. Their hard work and dedication have made this collection possible and have significantly advanced the field of engineering materials in extreme environments. We would also like to thank the reviewers for their constructive feedback and the editorial team for their support in bringing this Special Issue to fruition.

Looking ahead, we believe that this Special Issue will serve as a valuable resource for researchers, engineers, and students working in the field of engineering materials in extreme environments. It will provide a foundation for future research and development efforts aimed at improving the performance and reliability of materials in demanding applications. We hope that this collection will inspire further innovation and collaboration in this critical area.

**Yi Gong and Qi Tong**
*Guest Editors*

materials

Article

# Analysis of the Accelerator-Driven System Fuel Assembly during the Steam Generator Tube Rupture Accident

Di-Si Wang [1], Bo Liu [1], Sheng Yang [1], Bin Xi [1], Long Gu [2,3,4,*], Jin-Yang Li [2], Janne Wallenius [5] and You-Peng Zhang [1,*]

1 Institute of Modern Physics, Fudan University, Shanghai 200433, China; 20210200015@fudan.edu.cn (D.-S.W.); 20210200011@fudan.edu.cn (B.L.); 18210200009@fudan.edu.cn (S.Y.); 19110200027@fudan.edu.cn (B.X.)
2 Institute of Modern Physics, Chinese Academy of Sciences, Lanzhou 730000, China; lijinyang@impcas.ac.cn
3 University of Chinese Academy of Sciences, Beijing 100049, China
4 School of Nuclear Science and Technology, Lanzhou University, Lanzhou 730000, China
5 KTH, Division of Nuclear Engineering, Albanova University Centre, 10691 Stockholm, Sweden; janne@neutron.kth.se
* Correspondence: gulong@impcas.ac.cn (L.G.); zhangyp@fudan.edu.cn (Y.-P.Z.)

**Abstract:** China is developing an ADS (Accelerator-Driven System) research device named the China initiative accelerator-driven system (CiADS). When performing a safety analysis of this new proposed design, the core behavior during the steam generator tube rupture (SGTR) accident has to be investigated. The purpose of our research in this paper is to investigate the impact from different heating conditions and inlet steam contents on steam bubble and coolant temperature distributions in ADS fuel assemblies during a postulated SGTR accident by performing necessary computational fluid dynamics (CFD) simulations. In this research, the open source CFD calculation software OpenFOAM, together with the two-phase VOF (Volume of Fluid) model were used to simulate the steam bubble behavior in heavy liquid metal flow. The model was validated with experimental results published in the open literature. Based on our simulation results, it can be noticed that steam bubbles will accumulate at the periphery region of fuel assemblies, and the maximum temperature in fuel assembly will not overwhelm its working limit during the postulated SGTR accident when the steam content at assembly inlet is less than 15%.

**Keywords:** accelerator-driven system; heavy liquid metal; CFD simulation; two-phase flow

**Citation:** Wang, D.-S.; Liu, B.; Yang, S.; Xi, B.; Gu, L.; Li, J.-Y.; Wallenius, J.; Zhang, Y.-P. Analysis of the Accelerator-Driven System Fuel Assembly during the Steam Generator Tube Rupture Accident. *Materials* **2021**, *14*, 1818. https://doi.org/10.3390/ma14081818

Academic Editor: Sofoklis Makridis

Received: 19 February 2021
Accepted: 27 March 2021
Published: 7 April 2021



## 1. Introduction

Nuclear energy, as a type of clean energy, can effectively reduce carbon dioxide emissions and therefore the greenhouse effect. At present, the fourth-generation nuclear technology that can provide cleaner and safer nuclear power has attracted gradually more attention. In the current nuclear industry, nuclear waste disposal is considered to be one of the most significant problems [1]. During the International Atomic Energy Agency Conference, six advanced nuclear reactor systems were proposed as Generation-IV reactor designs, which were considered to be able to partially solve this problem. Lead-based fast reactors [2–4], as one of these reactor types, have been proved to be able to efficiently transmute long-life radioactive nuclides in nuclear waste thanks to their excellent inherent safety characteristics [5,6].

At present, plenty of countries have proposed their own lead-based fast reactors (LFR) and accelerator-driven subcritical systems (ADS) designs, as listed in Table 1.

ADS was considered one of the most promising devices to perform nuclear waste transmutation, which applies an accelerator to provide a high-energy and high-current proton beam, which bombards heavy nuclei to produce high-flux broad-spectrum spallation neutrons. This external neutron source can drive and maintain the continuous and stable operation of the subcritical core. The reactor can burn nuclear waste and convert long-life nuclides to short-life nuclides with sufficient margin to core failure thanks to its inherent safety characteristics. After more than 20 years of preliminary research, at the end of 2015, the China initiative accelerator-driven system (CiADS) project was approved. The CiADS project adopts the technical route of the combination of a "superconducting linear accelerator + high-power spallation target + subcritical reactor". Its conceptual design has been completed, and a series of key scientific and technical research is still ongoing.

**Table 1.** Coolants in typical lead-based fast reactors (LFRs) designs [6–9].

| Reactor | Coolant | Country |
|---|---|---|
| SVBR-100 | LBE | Russia |
| SSTAR | Pb | American |
| LSFR | LBE | Japan |
| CiADS | LBE | China |
| CLEAR-SR | LBE | China |
| MYRRHA | LBE | Belgium |
| SEALER | Pb | Sweden |

In the CiADS project, a lead-bismuth eutectic (LBE) was proposed as the primary coolant thanks to its good neutron economy and suitable thermophysical properties [10]. Its primary loop mainly consists of core, primary pumps and steam generators [11]. In each steam generator unit, a large number of pipes are installed. In the primary side of pipe wall, the primary coolant LBE works at a pressure slightly higher than the atmosphere pressure, while the water in the secondary side of pipe works at a much higher pressure aiming at a better heat transfer efficiency. This may cause a steam intrusion into the primary side or even into the core region of CiADS when the steam generator tube ruptures (SGTR) accident happens [12]. Therefore, a safety analysis towards the SGTR accident should be considered when designing the CiADS [13,14].

With the continuous development of high-performance computers and numerical calculation methods, computational fluid dynamics (CFD) has been widely used in validating various ADS designs [15]. Chen [16] studied the bundle in single phase and simulated the pin bundle blockage in a fuel assembly. Koloszar [17] used the CFD model to investigate the flow pattern in the core region of MYRRHA. Zhang [13] simulated the two-phase flow in a heavy liquid metal pool and assessed bubble-risen behaviors in it. Jeltsov [14] simulated the two-phase flow in a heat-exchanger and found the SGTR accident may be worse when the broken area is close to the primary pump.

OpenFOAM is an excellent CFD analysis tool based on the finite volume method and written with C++ [18]. Many researchers are using OpenFOAM to perform complex flow simulations thanks to its editable solver and flexibility in governing equation settings.

The VOF method proposed by Hirt and Nichols [19] in 1981 has the advantages of fewer iterations and a higher accuracy when performing two-phase simulations because of the application of the Euler–Euler multiphase model and the fluid volume setting method. Zhu [20] performed a systematic study on the formation of droplets in gas microchannels using the VOF method. Chen [21] and others used the VOF method to simulate the interaction process of the water–air interface. Li [22–24] carried out a numerical simulation of the movement characteristics of a single bubble in a gas–liquid two-phase flow under high pressure. VOF method was proved to be suitable for simulating the multiphase flow between a variety of immiscible fluids by setting the gas-phase content in each grid.

In recent CFD studies, the k-epsilon model, the k-omega model, the SST model and the LES method were introduced to predict the turbulence flow. The k-epsilon model and the k-omega model are RANS models, which can provide higher calculation efficiency, but less accuracy in some cases [25]. The SST model was reported to be able to provide high accuracy for a low-Reynolds flow [26]. The LES model could predict the fluid more accurately than RANS, but with a higher requirement for calculation resources [25]. The MYRRHA project [17] and Sugrue [18] recommended to use the k-epsilon model for two-phase LBE flow simulations. Therefore, we considered it to be suitable for the studies performed in this paper [27–30]. Waite [31] reported that the simulation results calculated with the k-epsilon model for the two-phase flow in rod bundles were accordant with experimental results. The standard k-epsilon model was then chosen for the turbulence simulations performed in this study.

In recent studies, the first-order upwind scheme, the second-order upwind scheme, the central difference scheme and the QUICK scheme were performed. The first-order upwind scheme is widely used in engineering for its efficiency, but it may be not accurate. The central difference scheme and the QUICK scheme may be unstable. The second-order upwind scheme has the intercept of second-order accuracy, but it still has the problem of false diffusion. According to the studies performed by other researchers [18,32], in this work, the second-order upwind convection scheme was adopted. Time integration was performed using the first-order method. The PIMPLE algorithm was used as the pressure-velocity coupling scheme, which is a combination of the PISO and the SIMPLE algorithm by OpenFOAM.

In this paper, the OpenFOAM based on the VOF method was adopted to study the two-phase flow under different inlet and heating conditions. In Section 2, the simulation model was created, including implementing the governing equations of VOF, meshing with ADS configurations and setting the coolant properties and boundary conditions. The validation of our simulation model was also included in this section. In Section 3, the simulation results of different inlet steam contents and heating conditions are summarized and discussed.

## 2. Materials and Methods

### 2.1. Governing Equations

In the VOF model, gas phase is considered to have the same velocity as the liquid phase. Among them, the mixed density and mixed dynamic viscosity are calculated as:

$$\rho = \alpha\rho_l + (1-\alpha)\rho_g \tag{1}$$

$$\mu = \alpha\mu_l + (1-\alpha)\mu_g \tag{2}$$

The governing equation of $\alpha$ is:

$$\frac{\partial\alpha}{\partial t} + \nabla\cdot(\alpha U) = 0 \tag{3}$$

The mass equation can be described as:

$$\frac{\partial\rho}{\partial t} + \nabla\cdot(\rho U) = 0 \tag{4}$$

The momentum equation can be described as:

$$\frac{\partial\rho U}{\partial t} + \nabla\cdot(\rho UU) - \nabla\cdot\tau = -\nabla p + \rho g + \sigma\kappa\nabla\alpha \tag{5}$$

The energy equation can be described as:

$$\frac{\partial\rho T}{\partial t} + \nabla\cdot(\rho UT) - \nabla\cdot\left(\frac{k_f}{c_p}\nabla T\right) = 0 \tag{6}$$

The equation of turbulence kinetic energy $k$ is:

$$\frac{\partial(\rho k)}{\partial t} + \nabla\cdot(\rho Uk) = \nabla\cdot\left[\left(\mu_m + \frac{\mu_t}{\sigma_k}\right)\nabla k\right] + G_k - \rho\varepsilon \tag{7}$$

The equation of turbulence kinetic energy dissipation epsilon is

$$\frac{\partial(\rho\varepsilon)}{\partial t} + \nabla\cdot(\rho U\varepsilon) = \nabla\cdot\left[\left(\mu_m + \frac{\mu_t}{\sigma_\varepsilon}\right)\nabla\varepsilon\right] + \frac{\varepsilon}{k}(C_{\varepsilon1}G_k - C_{\varepsilon2}\rho\varepsilon) \tag{8}$$

Then, Equations (7) and (8) are coupled by the following equations:

$$\mu_t = \rho C_\mu k^2/\varepsilon \tag{9}$$

$$G_k = \mu_t\nabla U\cdot\left[\nabla U + (\nabla U)^T\right] \tag{10}$$

where $\mu_m$ is the dynamic viscosity of laminar fluid; $\mu_t$ is the dynamic viscosity of turbulent fluid; $\sigma_k$ and $\sigma_\varepsilon$ represent diffusion Prandtl numbers ($\sigma_k = 1.0$ and $\sigma_\varepsilon = 1.3$); the model constants $C_\mu$, $C_{\varepsilon1}$ and $C_{\varepsilon2}$ are 0.09, 1.44 and 1.92, respectively.

### 2.2. Meshing the ADS Assembly

The ADS project adopts a subcritical reactor, whose thermal power is 8 MW. Fuel assemblies in the ADS subcritical core adopt a regular hexagonal outer tube encapsulation structure. The pin diameter is 6.55 mm and the fuel rod center distance is 9.17 mm. In order to prevent the fuel assembly from floating in the lead–bismuth coolant, the counterweight and the locking mechanism are installed at the lower end of the fuel assembly. The 3D structure of the fuel assembly is shown in Figure 1.

**Figure 1.** The 3D structure of the fuel assembly.

As shown in Figure 2, the flow channel in the fuel assemblies can be subdivided as internal channels, edge channels and corner channels based on their different heat transfer characteristics. The internal channel is heated by three adjacent one-sixth fuel rods, whose total heating area is a half fuel rod. The edge channel is heated by two-quarter fuel rods, whose total heating area is a half fuel rod, and one patch is an adiabatic plane. The corner channel is heated by a one-sixth fuel rod, and two patches are adiabatic planes. These differences in heating area and hydraulic diameters will affect the coolant flow patterns in the channels.

Figure 3 shows the grid division on the Z plane. The total number of cells is 5,478,900, the max skewness is 0.79, the mesh orthogonal quality is higher than 0.7 and the y-plus value is 30. To reduce the computational complexity, the space wire was neglected from the grid division and the grid therefore was a hexahedral grid. A coarse mesh with 4,812,000 cells and a fine mesh with 11,022,000 cells were built to compare with the 5,478,900 cells in this study. After given the same boundary condition, the simulation results with the 5,478,900 cells were reported to be close to those with the 11,022,000 cells as shown in Figure 4. The simulation was conducted using a Dell Server 7920, which has 80-core CPUs and 128GB RAM. For a simulation of a 10 s scenario, the computing time of meshes with 4,812,000 cells, 5,478,900 cells and 11,022,000 cells were 30.5 h, 42.3 h and 103.2 h, respectively. Therefore, considering the computational economy and accuracy, the mesh with 5,478,900 cells was chosen for this study.

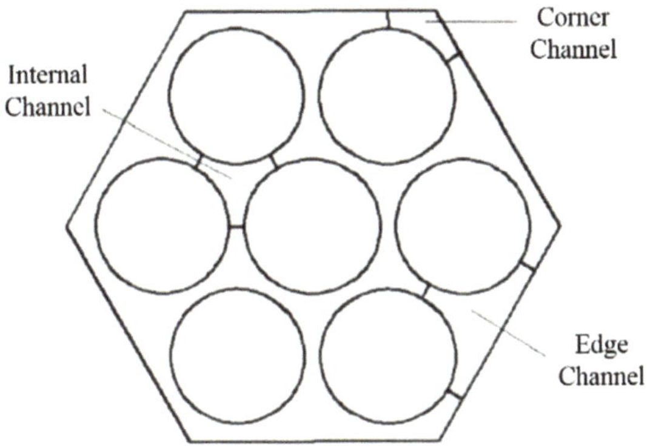


**Figure 2.** Fuel assembly subchannels.

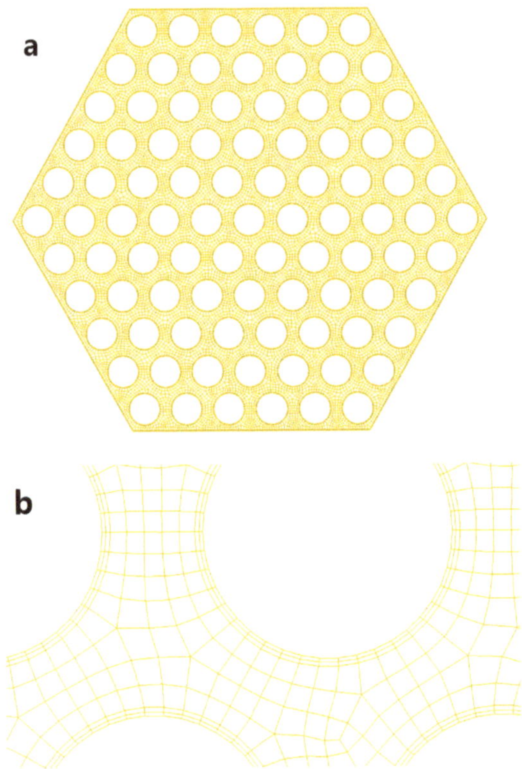


**Figure 3.** Grid division of an assembly (**a**) and boundary zones (**b**).

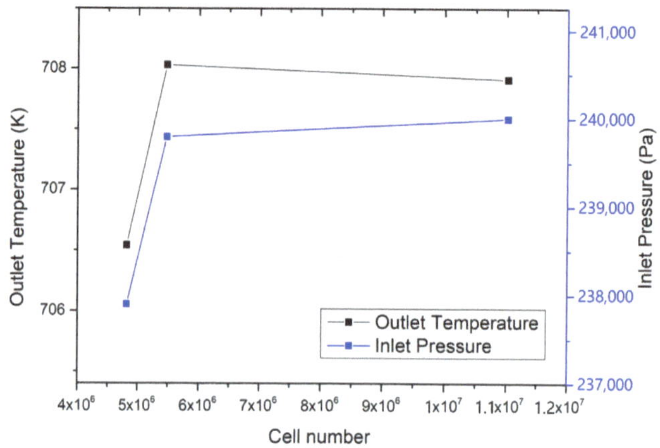


**Figure 4.** Grid convergence study.

### 2.3. Coolant Properties

The thermophysical parameters of lead–bismuth eutectic (LBE) and lead (Pb) coolant have been experimentally measured and fitted into correlations as listed in Table 2 [33].

**Table 2.** Properties of liquid lead–bismuth eutectic (LBE) and Pb.

| Properties | Liquid LBE | Liquid Pb |
|---|---|---|
| $\rho[kg/m^3]$ | $11096 - 1.3236T$ | $11367 - 1.1944T$ |
| $\mu[Pa \cdot s]$ | $4.94 \cdot 10^{-4} \cdot \exp(754.1/T)$ | $4.55 \cdot 10^{-4} \cdot \exp(1069/T)$ |
| $c_p\left[\frac{J}{kg \cdot K}\right]$ | $159.0 - 0.0272T + 7.12 \cdot 10^{-6}(T)^2$ | $161.5 - 0.0268T + 6.98 \cdot 10^{-6}(T)^2$ |
| $\lambda\left[\frac{W}{m \cdot K}\right]$ | $3.61 + 1.517 \cdot 10^{-2}T - 1.741 \cdot 10^{-6}(T)^2$ | $9.2 + 0.011T$ |

The steam properties obtained from the international standard IAPWS-IF97 steam table were defined into the OpenFOAM material database [34]. These steam properties were reported to be valid in the temperature region of 273.15 K–2273.15 K and for pressure lower than 10 MPa.

### 2.4. Boundary Conditions Setting

Figure 5 shows the boundary condition settings of the fuel assembly. The inlet velocity of fluid was defined as 0.2 m/s with the direction vertical to the patch. The wall was defined as a no-slip wall [35]. The outlet flow condition was defined as a pressure outlet equal to the atmosphere pressure. The transient mode was adopted to simulate bubble behaviors in the fuel assembly.

To simulate the distribution of bubbles in the fuel assembly under different boundary conditions, simulations were divided into 4 cases as listed in Table 3.

### 2.5. Verification of Our Simulation Model

The migration of steam bubbles in heavy liquid metal is complicated. In this process, bubbles are subjected to buoyancy, surface tension, drag force and other effects. The floating process cannot be optically observed, nor can it be observed by X-ray penetration due to the high density and opacity of the heavy liquid metal. To verify our simulation model, we referred to the experimental results obtained with an experimental device [13], in which high-speed gas was injected into the water tank to form a two-phase flow and recorded by the high-speed camera. Figure 6 shows the set-up of this experimental device, which includes a gas injection system, a water tank, a high-speed camera, sensors and measurement systems. The high-pressure air produced by the compressor was injected into the water tank through a pipe. The diameter of the pipe nozzle is 4 mm and it was located at 5 mm under the water. The water tank is a glass cuboid with a length of 250 mm, a width of 250 mm, and a height of 800 mm. The high-speed camera is used to observe the bubble shape and the depth of injection. During the experiment, the gas injection velocity was set as 22.12 m/s, 26.54 m/s, 33.18 m/s, 35.39 m/s, 44.24 m/s, 55.30 m/s, 66.36 m/s, 77.42 m/s, 88.48 m/s and 99.54 m/s, respectively.

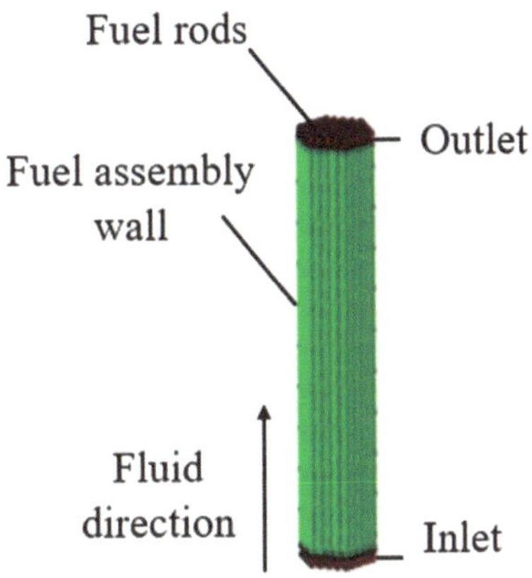


**Figure 5.** Boundary condition settings.

**Table 3.** The setting of different cases.

| Case Number | Coolant | Heating Boundary Condition |
|---|---|---|
| Case 1 | LBE | Fixed Heat Flux value |
| Case 2 | LBE | Fixed Wall Temperature |
| Case 3 | Pb | Fixed Heat Flux value |
| Case 4 | Pb | Fixed Wall Temperature |

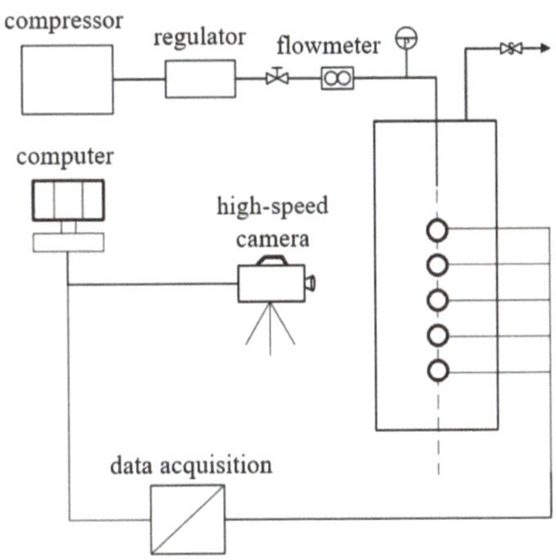


**Figure 6.** Schematic diagram of the experimental apparatus.

In order to model this experiment into OpenFOAM, we first constructed a model of the water tank. The surrounding walls and the bottom of the tank were defined as no-slip walls in our simulation model. The top of the tank was defined as an outflow patch at the ambient pressure. The nozzle was modeled in the center of the top patch. The computational area was divided into 250, 250 and 800 meshes in the X, Y and Z directions. The model was constructed based on the geometrical details of the experiment tank.

Figure 7 shows the results from the experiments and the simulations. As can be seen from Figure 7, the experimental and simulation results are in good agreement. The maximum relative error (approximately 7.7%) occurred at the velocity of 22.12 m/s, which is considered to be mainly caused by the small denominator. The absolute differences between the experimental and the simulated results were reported to be less than 2 mm. Based on the results stated above, the VOF model was considered to be suitable for simulations performed in the following sections.

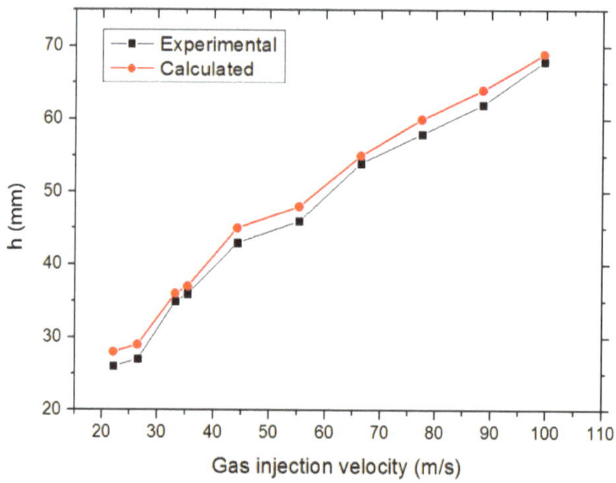


**Figure 7.** Bubble penetration depth for the gas injection into water.

## 3. Simulations under Different Boundary Conditions

### *3.1. Temperature Distributions at the Middle Section and the Outlet Section*

Figures 8 and 9 show the maximum temperatures at the middle section in case 1 and case 3. The middle section was set in parallel with the inlet section and allocated at the center of the fuel assembly. When the coolant was pure LBE coolant, the maximum temperature at the middle section was reported to be 646 K. When the coolant was pure liquid Pb, the maximum temperature at the middle section was 651 K. This means that the designed temperature of the Pb coolant reactor was slightly higher than that of the LBE coolant. When the inlet steam content increases from 1 to 15%, the temperature at the middle section will slowly increase. When the inlet steam content was less than 15%, the temperature in the reactor fuel assembly was reported to be lower than its working limit. When the inlet steam content exceeds 15%, the temperature will rise significantly and the temperature fluctuations will also become larger.

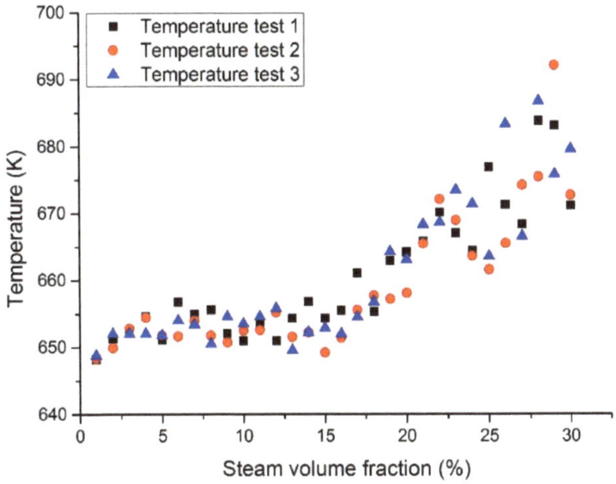


**Figure 8.** Maximum LBE temperature at the assembly middle section with a fixed heat flux (case 1).

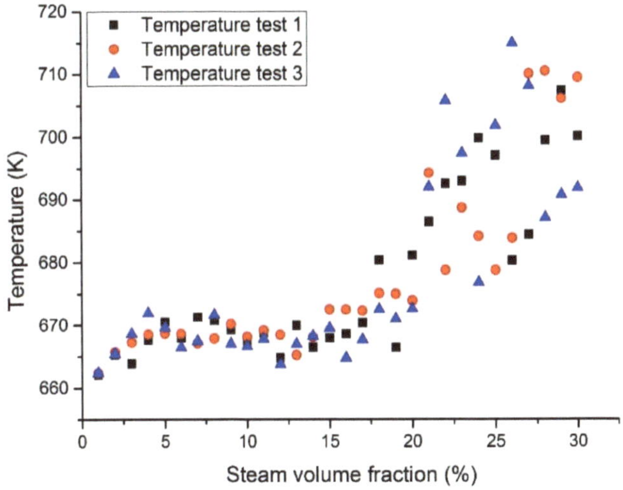


**Figure 9.** Maximum Pb temperature at the assembly middle section with a fixed heat flux (case 3).

Figures 10 and 11 show the maximum temperatures at the outlet section in case 1 and case 3. According to the three test results, the maximum temperature of liquid Pb was higher than that of the lead–bismuth coolant with the same inlet steam content. When a large number of bubbles enter, the temperature fluctuation of the Pb coolant is greater than that of the lead–bismuth coolant. When the inlet steam content is less than 15%, the calculated fluid temperature is lower than its working limit. It was reported that the fuel bundle may burn up when the inlet steam content exceeds 15% due to the high cladding temperature incurred by insufficient cooling.

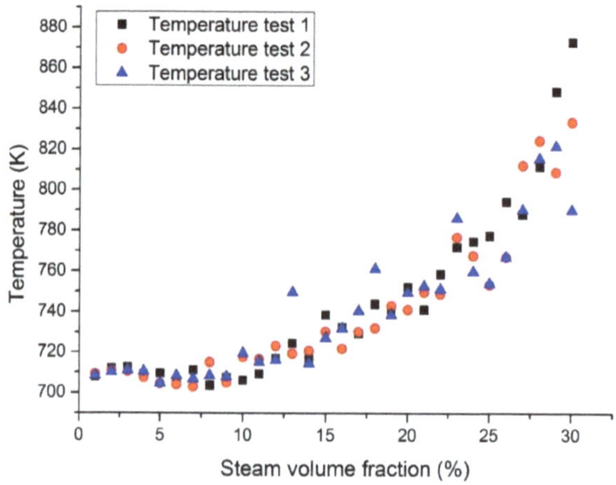


**Figure 10.** Maximum LBE temperature at the assembly outlet section with a fixed heat flux (case 1).

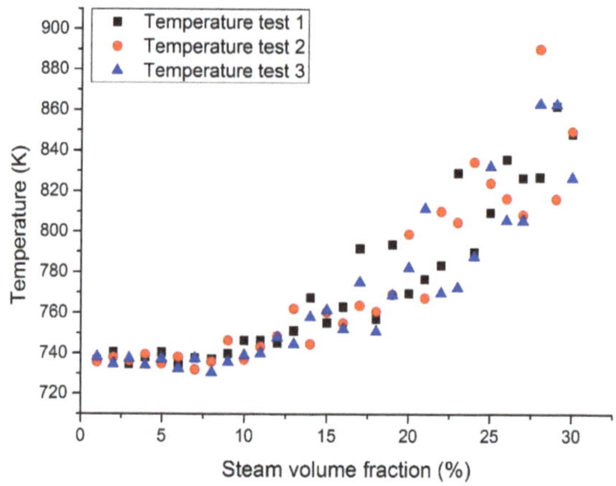


**Figure 11.** Maximum Pb temperature at the assembly outlet section with a fixed heat flux (case 3).

## 3.2. Bubble Gathering

Figure 12 shows the distribution of the gas phase when the inlet steam content is 10% in each case. Figure 13 shows the bubble accumulation at the assembly outlet section with a fixed tube wall temperature. It can be noticed that the steam bubbles will accumulate at the periphery regions. The probability of the largest bubble appearing in the corner channel exceeds 50%. This phenomenon may cause fuel rods to operate under risky conditions. The bubble accumulation may be caused by the larger fluid velocity in the internal channel than in the periphery channel.

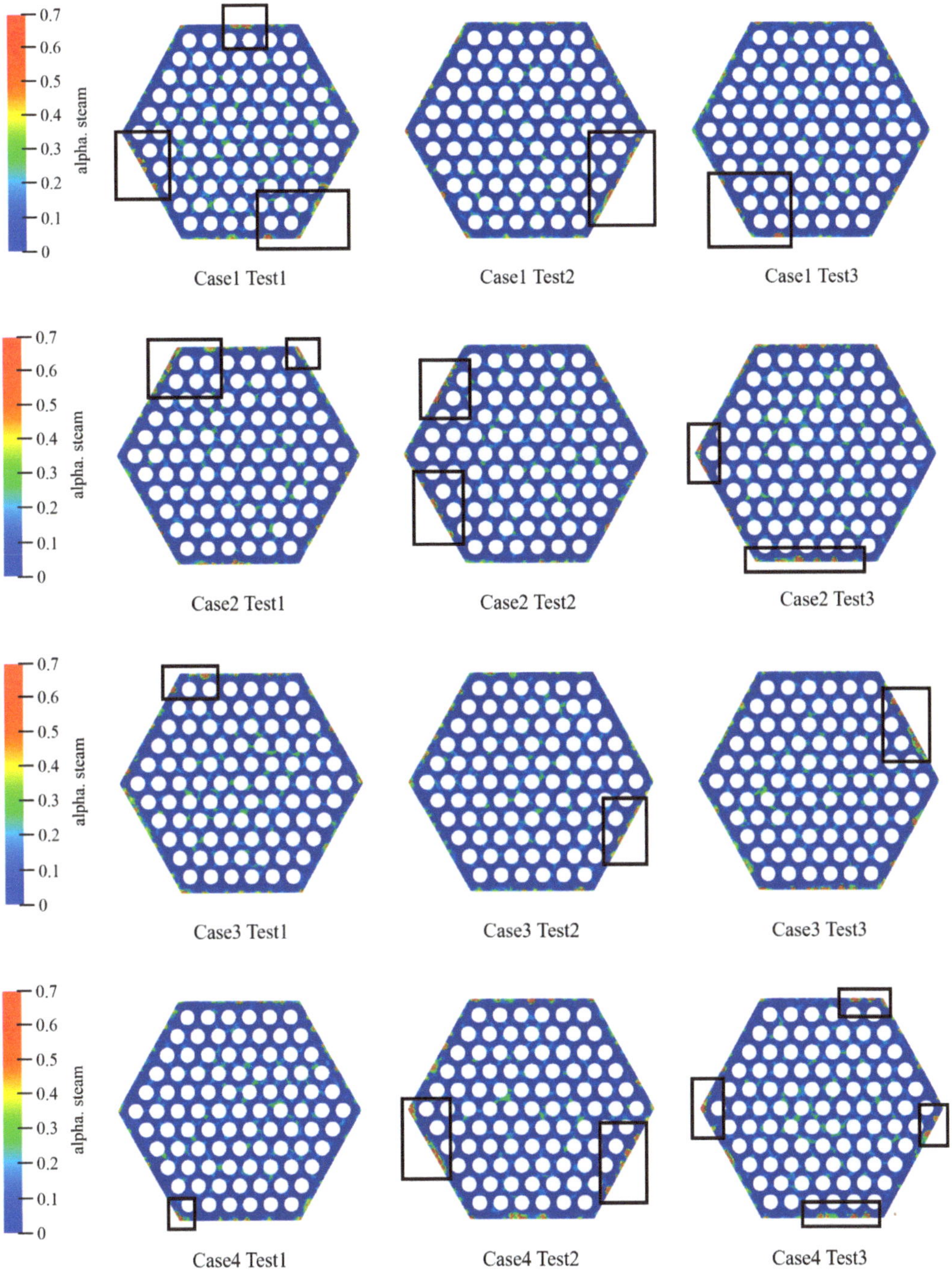


**Figure 12.** Bubble distribution in the cases 1 to 4 with 10% inlet steam content. The α steam stands for the steam content.

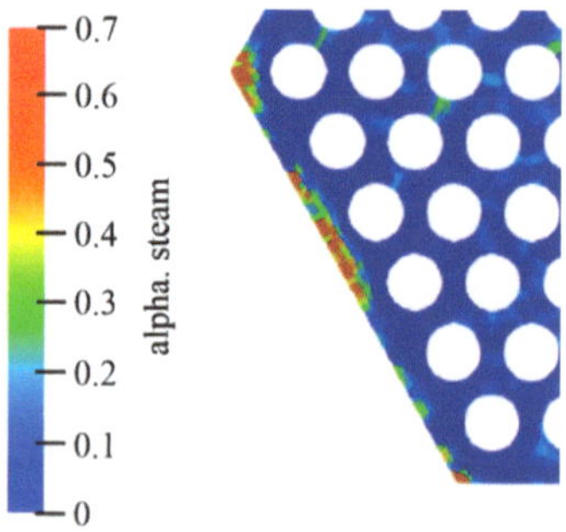


**Figure 13.** Steam bubbles at the periphery regions.

### *3.3. Outlet Section Velocity*

Figures 14–17 show the maximum velocity at the outlet section. It can be seen from these figures that the maximum velocity gradually increases with the increase of the inlet steam content. Together with the outlet section temperatures, it can be noticed that the temperature has little effect on velocity, mainly because the steam content plays a leading role in velocity. Due to the steam expansion, the two-phase flow velocity will rise with the inlet steam content.

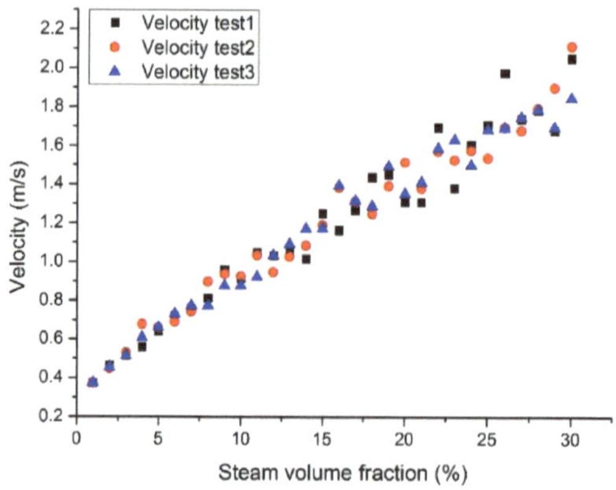


**Figure 14.** Maximum LBE velocity at the assembly outlet section with a fixed heat flux (case 1).

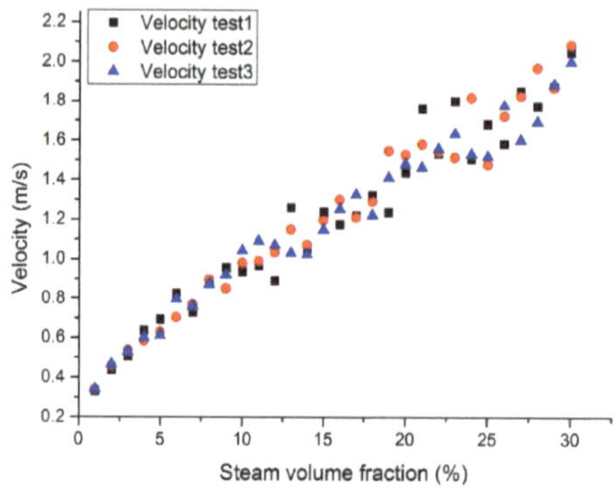


**Figure 15.** Maximum LBE velocity at the assembly outlet section with a fixed tube wall temperature (case 2).

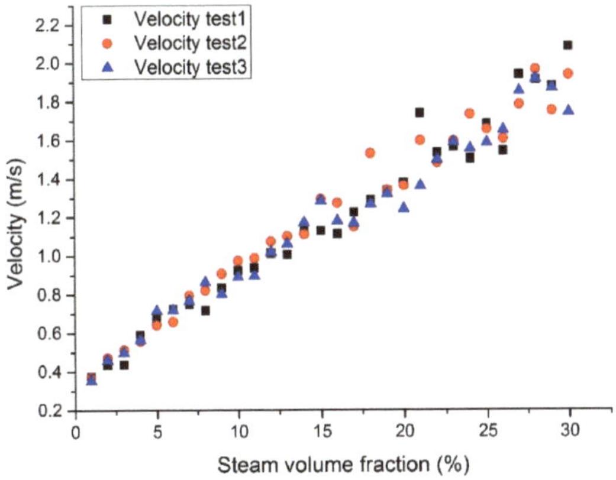


**Figure 16.** Maximum Pb velocity at the assembly outlet section with a fixed heat flux (case 3).

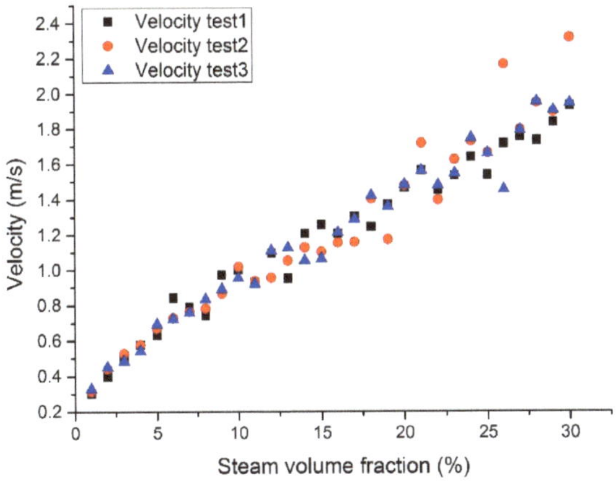


**Figure 17.** Maximum Pb velocity at the assembly outlet section with a fixed tube wall temperature (case 4).

### 3.4. Maximum Steam Volumetric Fraction

Figures 18–21 show the maximum grid steam volumetric content at the middle section and the outlet section. The steam content at the middle section is lower than that at the outlet section. The steam content in the grid increases significantly from 5 to 15%. When the inlet steam content is higher than 15%, the floating speed slows down because of the occupation of steam bubbles in the upper region. When the inlet steam content is less than 2%, the grid steam content at the middle section and the outlet section are basically the same. When the inlet steam content increases from 5 to 20%, the grid steam content at the middle section and the outlet section are significantly different. It was reported that the bubbles would break and merge together under this condition, which may cause fluctuations of fuel rods temperature and therefore thermal fatigue. When the inlet steam content exceeds 20%, the difference becomes smaller.

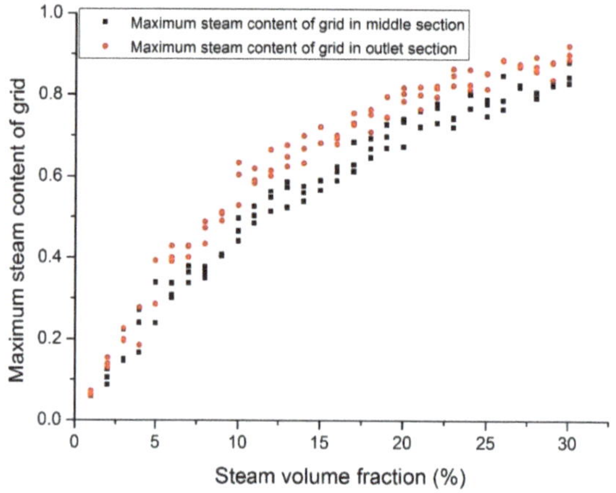


**Figure 18.** Maximum steam content at the assembly outlet section (case 1).

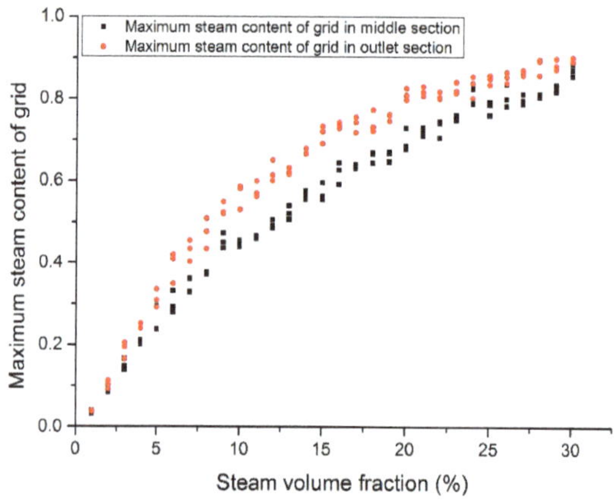


**Figure 19.** Maximum steam content at the assembly outlet section (case 2).

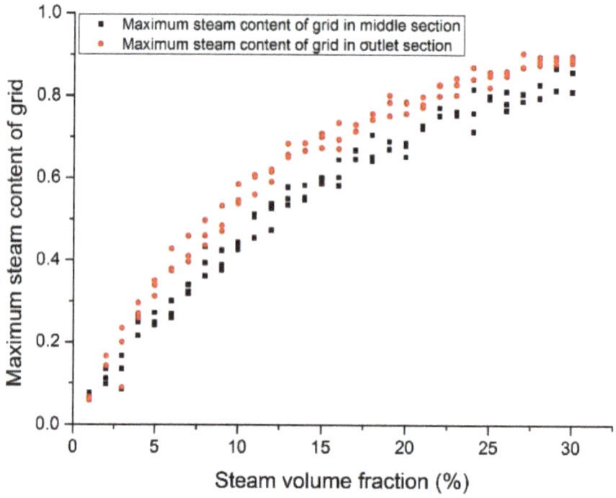


**Figure 20.** Maximum steam content at the assembly outlet section (case 3).

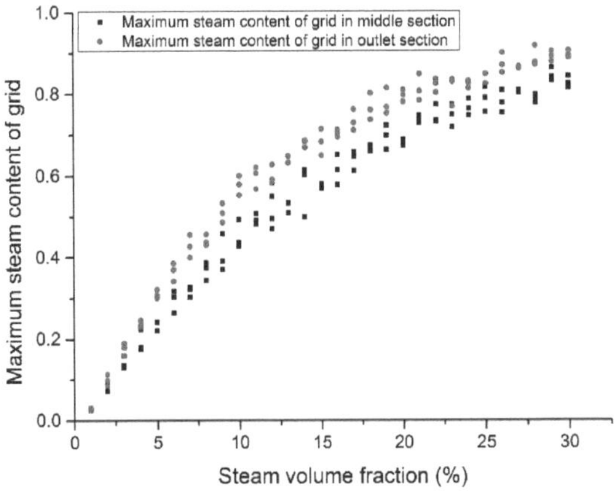


**Figure 21.** Maximum steam content at the assembly outlet section (case 4).

## 4. Conclusions

In this paper, the open source CFD software OpenFOAM was adopted to simulate the two-phase heavy liquid metal flow. Based on our simulation results, the following points can be preliminarily concluded:

(1) When the steam content is lower than 15%, the coolant temperature rises slowly. The fuel rod would operate within its designed limit. A significant rise of coolant temperature may be noticed after the inlet steam content exceeds 15%.
(2) The temperature fluctuation in liquid Pb (lead) due to the steam bubble intrusion was reported to be slightly higher than that in LBE coolant.
(3) The temperature fluctuation is significant when the inlet steam content is higher than 15%, which may cause the failure of fuel assemblies. It was reported that this significant phenomenon needs to be carefully investigated.
(4) Steam bubbles tend to accumulate at the periphery regions, especially in the corner channels. This phenomenon confirms the necessity of constructing corresponding experimental facilities in the future.
(5) The outlet fluid velocity and the steam content were reported to be mainly affected by the inlet steam content.

**Author Contributions:** Conceptualization, D.-S.W., B.L., S.Y., B.X. and Y.-P.Z.; formal analysis, D.-S.W., B.L., S.Y., L.G., J.-Y.L. and J.W.; investigation, D.-S.W., B.L., S.Y. and B.X.; methodology, D.-S.W., B.L. and Y.-P.Z.; software, D.-S.W. and B.L.; writing—original draft, D.-S.W. and S.Y.; writing—review and editing, B.L., B.X. and Y.-P.Z. All authors have read and agreed to the published version of the manuscript.

**Funding:** This work was supported partly by the Ministry of Science and Technology of the People's Republic of China (approved number: 2020YFB1902100). This research was financially supported by the Shanghai Municipal Commission of Economy and Informatization (approved number: GYQJ-2018-2-02).

**Institutional Review Board Statement:** Not applicable.

**Informed Consent Statement:** Not applicable.

**Data Availability Statement:** Data sharing not applicable.

**Conflicts of Interest:** The authors declare no conflict of interest.

## Nomenclature

| Symbol | Explanation |
|---|---|
| $\rho$, $\rho_l$ and $\rho_g$ | Mixed Fluid Density, Liquid Density and Gas Density (kg·m$^{-3}$) |
| $\alpha$ | volume fraction of fluid |
| $\mu$, $\mu_l$ and $\mu_g$ | mixed fluid dynamic viscosity, liquid dynamic viscosity and gas dynamic viscosity (Pa·s) |
| t | time (s) |
| U | Gas phase and liquid phase mixed velocity (m·s$^{-1}$) |
| $\tau$ | viscous stress (Pa) |
| p | pressure (Pa) |
| g | gravity (m·s$^{-2}$) |
| $\sigma$ | surface tension coefficient (N·m$^{-1}$) |
| $\kappa$ | curvature (m$^{-1}$) |
| T | temperature (K) |
| $k_f$ | fluid heat transfer coefficient (W·m$^{-2}$·K$^{-1}$) |
| $c_p$ | specific heat capacity at constant pressure (J·kg$^{-1}$·K$^{-1}$) |
| $\lambda$ | thermal conductivity (W·m$^{-1}$·K$^{-1}$) |

## References


1. Gudowski, W. Accelerator-driven transmutation projects. The importance of nuclear physics research for waste transmutation. *Nucl. Phys. A* **1999**, *654*, 436c–457c. [CrossRef]
2. Wallenius, J.; Qvist, S.; Mickus, I.; Bortot, S.; Szakalos, P.; Ejenstam, J. Design of SEALER, a very small lead-cooled reactor for commercial power production in off-grid applications. *Nucl. Eng. Des.* **2018**, *338*, 23–33. [CrossRef]
3. Takahashi, M.; Uchida, S.; Kasahara, Y. Design study on reactor structure of Pb–Bi-cooled direct contact boiling water fast reactor (PBWFR). *Prog. Nucl. Energy* **2008**, *50*, 197–205. [CrossRef]
4. Wu, Y.C.; Bai, Y.Q.; Song, Y.; Huang, Q.Y.; Zhao, Z.M.; Hu, L.Q. Development strategy and conceptual design of China Lead-based Research Reactor. *Ann. Nucl. Energy* **2016**, *87*, 511–516. [CrossRef]
5. Peng, T.J.; Lang, M.G.; Zhou, Z.W.; Gu, L. Accident analysis of tungsten target coupled with ADS core. *Int. J. Hydrog. Energy* **2016**, *41*, 7059–7068. [CrossRef]
6. Peng, T.; Zhou, Z.; Xiao, S.; Sheng, X.; Gu, L. Thermal–hydraulic modeling and transient analysis of helium-cooled tungsten target coupled with ADS core. *Ann. Nucl. Energy* **2016**, *87*, 612–620. [CrossRef]
7. Romanets, Y.; Abderrahim, H.A.; De Bruyn, D.; Dagan, R.; Goncalves, I.; Maschek, W.; Rimpault, G.; Struwe, D.; Van den Eynde, G.; Vaz, P.; et al. Xt-Ads: Neutronics, Shielding, and Radiation Damage Calculations. *Nucl. Technol.* **2009**, *168*, 537–541. [CrossRef]
8. Maes, D. Mechanical design of the small-scale experimental ADS: MYRRHA. *Energy Convers Manag.* **2006**, *47*, 2710–2723. [CrossRef]
9. Wallenius, J.; Bortot, S. A small lead-cooled reactor with improved Am-burning and non-proliferation characteristics. *Ann. Nucl. Energy* **2018**, *122*, 193–200. [CrossRef]
10. Subbotin, V.I.; Arnol'dov, M.N.; Kozlov, F.A.; Shimkevich, A.L. Liquid-metal coolants for nuclear power. *At. Energy* **2002**, *92*, 29–40. [CrossRef]
11. Mangialardo, A.; Borreani, W.; Lomonaco, G.; Magugliani, F. Numerical investigation on a jet pump evolving liquid lead for GEN-IV reactors. *Nucl. Eng. Des.* **2014**, *280*, 608–618. [CrossRef]
12. Wang, S.; Flad, M.; Maschek, W.; Agostini, P.; Pellini, D.; Bandini, G.; Suzuki, T.; Morita, K. Evaluation of a steam generator tube rupture accident in an accelerator driven system with lead cooling. *Prog. Nucl. Energy* **2008**, *50*, 363–369. [CrossRef]
13. Zhang, C.D.; Sa, R.Y.; Zhou, D.N.; Zhang, G.Y.; Wang, L. Effects of failure location and pressure on the core voiding under SGTR accident in a LBE-cooled fast reactor. *Int. J. Heat Mass Transf.* **2019**, *141*, 940–948. [CrossRef]
14. Jeltsov, M.; Villanueva, W.; Kudinov, P. Steam generator leakage in lead cooled fast reactors: Modeling of void transport to the core. *Nucl. Eng. Des.* **2018**, *328*, 255–265. [CrossRef]
15. Cheng, X.; Tak, N.I. CFD analysis of thermal–hydraulic behavior of heavy liquid metals in sub-channels. *Nucl. Eng. Des.* **2006**, *236*, 1874–1885. [CrossRef]
16. Chen, X.N.; Li, R.; Rineiski, A.; Jager, W. Macroscopic pin bundle model and its blockage simulations. *Energy Convers. Manag.* **2015**, *91*, 93–100. [CrossRef]
17. Koloszar, L.; Buckingham, S.; Planquart, P.; Keijers, S. MyrrhaFoam: A CFD model for the study of the thermal hydraulic behavior of MYRRHA. *Nucl. Eng. Des.* **2017**, *312*, 256–265. [CrossRef]
18. Sugrue, R.; Magolan, B.; Lubchenko, N.; Baglietto, E. Assessment of a simplified set of momentum closure relations for low volume fraction regimes in STAR-CCM plus and OpenFOAM. *Ann. Nucl. Energy* **2017**, *110*, 79–87. [CrossRef]
19. Hirt, C.W.; Nichols, B.D. Volume of fluid (VOF) method for the dynamics of free boundaries. *J. Comput. Phys.* **1981**, *39*, 201–225. [CrossRef]

20. Zhu, X.; Sui, P.C.; Djilali, N. Three-dimensional numerical simulations of water droplet dynamics in a PEMFC gas channel. *J. Power Sources* **2008**, *181*, 101–115. [CrossRef]
21. Chen, L.; Li, Y.G. A numerical method for two-phase flows with an interface. *Environ. Model. Softw.* **1998**, *13*, 247–255. [CrossRef]
22. Li, Y.; Zhang, J.P.; Fan, L.S. Discrete-phase simulation of single bubble rise behavior at elevated pressures in a bubble column. *Chem. Eng. Sci.* **2000**, *55*, 4597–4609. [CrossRef]
23. Li, Y.; Yang, G.Q.; Zhang, J.P.; Fan, L.S. Numerical studies of bubble formation dynamics in gas–liquid–solid fluidization at high pressures. *Powder Technol.* **2001**, *116*, 246–260. [CrossRef]
24. Yang, G.Q.; Du, B.; Fan, L.S. Bubble formation and dynamics in gas–liquid–solid fluidization—A review. *Chem. Eng. Sci.* **2007**, *62*, 2–27. [CrossRef]
25. Zhu, Z.Y.; Qin, B.; Li, S.H.; Liu, Y.G.; Li, X.; Cui, P.Z.; Wang, Y.L.; Gao, J. Multi-dimensional analysis of turbulence models for immiscible liquid-liquid mixing in stirred tank based on numerical simulation. *Sep. Sci. Technol.* **2021**, *56*, 411–424. [CrossRef]
26. Jeong, Y.S.; Kim, J.Y.; Bang, I.C. Enhanced heat transfer and reduced pressure loss with U-pattern of helical wire spacer arrangement for liquid metal cooled-fast reactor fuel assembly. *Ann. Nucl. Energy* **2020**, *135*, 106971. [CrossRef]
27. Chen, R.; Tian, M.; Chen, S.; Tian, W.; Su, G.H.; Qiu, S. Three dimensional thermal hydraulic characteristic analysis of reactor core based on porous media method. *Ann. Nucl. Energy* **2017**, *104*, 178–190. [CrossRef]
28. Ali, M.Y.; Wu, G.W.; Liu, S.Y.; Jin, M.; Zhao, Z.M.; Wu, Y.C. CFD Analysis of Thermal Stratification under PLOFA transient in CLEAR-S. *Prog. Nucl. Energy* **2019**, *115*, 21–29. [CrossRef]
29. Niu, F.L.; Zhuo, W.Q.; Cai, J.C.; Su, X.W.; Hu, Y.Q.; Zhao, Y.G.; Ma, T.F.; Wang, Y.Y.; Yu, Y. The studies of mixing and thermal stratification in SMR containments. *Nucl. Eng. Des.* **2016**, *298*, 14–24. [CrossRef]
30. Xia, Z.H.; Lu, D.G.; Tang, J.X.; Liu, Y.Z.; Yang, J.; Feng, L.; Ye, S.S.; Guo, Z.X.; Zhang, Y.H. Three-dimensional thermal hydraulic transient calculation of coupled cold and hot sodium pools under a loss of feedwater accident in the China experimental fast reactor. *Ann. Nucl. Energy* **2020**, *139*. [CrossRef]
31. Waite, B.M.; Prasser, H.M.; Podowski, M.Z. Computational and experimental analysis of gas/liquid two-phase flow in rod bundles with mixing-vane spacer grids. *Nucl. Eng. Des.* **2020**, *360*. [CrossRef]
32. Zeng, Q.Y.; Cai, J.J. Three-dimension simulation of bubble behavior under nonlinear oscillation. *Ann. Nucl. Energy* **2014**, *63*, 680–690. [CrossRef]
33. OCDE; AEN. *Handbook on Lead-Bismuth Eutectic Alloy and Lead Properties, Materials Compatibility, Thermalhydraulics and Technologies*; OECD Publishing: Paris, France, 2015.
34. Wagner, W.; Cooper, J.R.; Dittmann, A.; Kijima, J.; Kretzschmar, H.J.; Kruse, A.; Mares, R.; Oguchi, K.; Sato, H.; Stocker, I.; et al. The IAPWS industrial formulation 1997 for the thermodynamic properties of water and steam. *J. Eng. Gas. Turb. Power* **2000**, *122*, 150–182. [CrossRef]
35. Zeng-fang, G.; Tao, Z.; Yun-qing, B.; Yong, S. Thermal-hydraulic Analysis in Wire-wrapped Fuel Assembly for China Lead-based Research Reactor. *At. Energy Sci. Technol.* **2015**, *49*. [CrossRef]

*materials*

MDPI

*Article*

# Crack Growth Simulation of Functionally Graded Materials Based on Improved Bond-Based Peridynamic Model

**Liyi Min [1,2], Qiwen Liu [1,2,*] and Lisheng Liu [1,2,3]**

[1] Hubei Key Laboratory of Theory and Application of Advanced Materials Mechanics, Wuhan University of Technology, Wuhan 430070, China; liyi_min@whut.edu.cn (L.M.); Liulish@whut.edu.cn (L.L.)

[2] Department of Engineering Structure and Mechanics, Wuhan University of Technology, Wuhan 430070, China

[3] State Key Laboratory of Advanced Technology for Materials Synthesis and Processing, Wuhan University of Technology, Wuhan 430070, China

* Correspondence: qiwen_liu@whut.edu.cn; Tel.: +86-139-7153-8711

**Abstract:** Functionally graded materials (FGMs) are widely used in the aerospace industry, especially for the thermal protection shields of aerospace vehicles. Studies show that the initiation and expansion of micro-cracks are important factors that adversely affect the service life of these shields. Based on the peridynamic theory of bonds, an improved peridynamic model is proposed in the present study for FGMs. In the proposed model, integral equivalence is applied to calculate the required material parameters. Obtained results reveal that this method can better reflect the gradient change of material properties.

**Keywords:** functionally graded materials; peridynamic method; integral equivalence; crack growth simulation

**Citation:** Min, L.; Liu, Q.; Liu, L. Crack Growth Simulation of Functionally Graded Materials Based on Improved Bond-Based Peridynamic Model. *Materials* **2021**, *14*, 3032. https://doi.org/10.3390/ma14113032

Academic Editors: Yi Gong and Qi Tong

Received: 27 April 2021
Accepted: 30 May 2021
Published: 2 June 2021



## 1. Introduction

With the development of science and technology in the past few decades, the aerospace industry has developed rapidly. Studies show that during the entire mission of a spacecraft, the materials used may face a very harsh working environment. Accordingly, requirements for the constituent materials of the spacecraft have become increasingly stringent. In this regard, scholars proposed the concept of functionally graded materials (FGMs) made of metals and ceramics. Further investigations reveal that since the composition of these materials changes continuously with spatial position, the material properties on the macroscopic level also change accordingly, thereby eliminating the interface layer between different materials. This characteristic allows FGMs to simultaneously exert the excellent mechanical properties of metals and the high-temperature resistance of ceramics. In practical applications, FGM failures mainly originate from defects in the material and crack propagation. Therefore, it is of extreme importance to investigate the crack propagation phenomenon in FGMs. However, since an FGM is a heterogeneous substance, the material composition may vary spatially. Accordingly, it is an enormous challenge to analyze crack propagation in FGMs.

In the classical continuum theory, spatial partial differential equations are used to describe the material deformation. However, crack propagation is a typical discontinuity problem such that spatial partial derivatives cannot be applied to this problem. Considering the computational difficulties of mathematical methods, it is generally assumed that the FGM is an isotropic substance. Literature review indicates that only a few investigations have been conducted so far on crack propagation in anisotropic FGMs. In this regard, Erdogan et al. systematically studied the static fracture problem of heterogeneous materials [1,2]. Moreover, Jin et al. analyzed the crack propagation problem in homogeneous and FGMs and found that as long as the material parameters were continuous and differentiable, singular fields of the crack tip in these materials were the same [3]. Marur et al. carried out experiments and performed numerical simulations to study the interface fracture of bimaterials and FGMs under different impact loads [4]. Kawasaki et al. applied the controllable burner heating method to simulate the real environment and studied the thermal fracture behavior of FGMs. The obtained results showed that orthogonal cracks form on the top surface of the ceramic layer during the cooling process. Moreover, the gradient layer forms transverse cracks and these cracks propagate, merge, and eventually peel the coating off [5]. Considering the difficulties and high costs of the crack growth experiment in FGMs, it is important to investigate this problem through numerical methods. Meanwhile, conventional experimental methods can hardly reveal the whole cracking process in FGMs, including crack initiation, crack propagation and complete fracture. Currently, different numerical methods, including the finite element method, extended finite element method, boundary element method and meshless method, can be applied to study the FGM crack growth problem. Nishioka et al. presented the concept of mixed-phase simulation with fracture-path prediction mode, in which the crack propagation path is simulated by using the proposed automatic moving finite element method, and the simulation results were in good agreement with the experimental results [6]. Fabbrocino et al. proposed a new methodology to predict dynamic crack propagation under generalized loading conditions, and a numerical model of permission was established based on this method. This model can be used to predict the geometric changes caused by the discontinuous evolution of existing materials; the accuracy of the model was verified by comparison to experimental results. Sensitivity analyses in terms of mesh dependence and time required for the solving procedure were also discussed [7]. Dirik et al. developed a mesh-independent computational algorithm and incorporated this into a commercial finite element software (Abaqus). For verification purposes, experimental crack path trajectories and fatigue life data available in the open literature were compared with computational results [8].

In the classical finite element method, the singularity problem of the spatial derivative is considered as a typical discontinuity problem of crack propagation. Moreover, in the extended finite element method, additional functions are introduced to deal with the crack problem. However, this method has limitations for complex cracks such as multiple cracks and crack bifurcation. Further investigations reveal that the boundary element method has its own shortcomings. More specifically, its scope of action requires a basic solution of the corresponding differential operator as a major premise, which is a challenging prerequisite for heterogeneous materials. Furthermore, the meshless method requires a kernel function, and it also has the limitation of the spatial derivative problem. Recently, Silling proposed the peridynamic (PD) method [9] as a vigorous scheme to solve discontinuous problems, which has attracted many scholars worldwide.

Unlike the finite element method, the PD method does not use partial differential equations, but integral equations. Meanwhile, the material damage is included in the constitutive of the PD method. Accordingly, this method can be applied to simulate the crack propagation problem in the material without the need to set a criterion. Bobaru et al. performed a peridynamics analysis on the dynamic crack growth of brittle materials under different convergence conditions such as m-convergence and $\delta$-convergence and verified the calculated results through experimentation. Accordingly, it was found that the peridynamics method is an effective scheme for simulating the dynamic crack growth problem [10]. Ha et al. used the bond-based PD theory to study the influence of the pre-crack angle on the crack growth of rock-like materials containing pre-crack subjected to uniaxial compressions. The obtained results proved the effectiveness of the PD method [11]. Cheng et al. proposed a near-field dynamic model for FGMs and studied the effects of different parameters, including the material gradient, elastic wave, impact load size and contact time on the fracture behavior of FGMs [12]. Tan et al. proposed a complete thermo-mechanical coupling peridynamics model with a surface correction near cracks, and then applied this model to perform thermodynamic simulations of the FGM plate with thermal insulation cracks [13].

Based on the performed literature survey, we intended to apply the PD theory and propose an improved bond-based peridynamics model. The proposed model will be used to simulate the crack growth of FGMs with the material discretized into scattered points. Therefore, the material properties of the bonds between different particles should be processed and calculated. It should be indicated that the conventional method to do so is to use the average value for processing. However, the material properties of FGMs are not uniform so the average solution method cannot reflect the internal properties of FGMs accurately. The proposed method is expected to reflect the variations of the properties in functionally graded materials.

## 2. Materials and Methods

### 2.1. Bond-Based Peridynamic Basic Theory

Figure 1 shows that in the bond-based PD theory, the object is discretized into several material points, where each material point x only interacts with other material points in the range of $\delta$. This interaction is represented by the bond $\xi = x' - x$. It is worth noting that the interaction between two material points is equal in magnitude and opposite in direction. At a certain time, $t$, this motion can be mathematically expressed in the form below:

$$\rho(x)\ddot{u}(x,t) = \int_{H_x} f(x',x,t)dV_{x'} + b(x,t) \tag{1}$$

where $f$ and $u$ denote the force function between material points and the displacement of the material point at a certain moment, respectively. Moreover, $H_x$, $\rho$ and $b$ are the neighborhood range, material density and force on the material point $x$, respectively. $V_{x'}$ is the volume of other points in the neighborhood of point $x$.

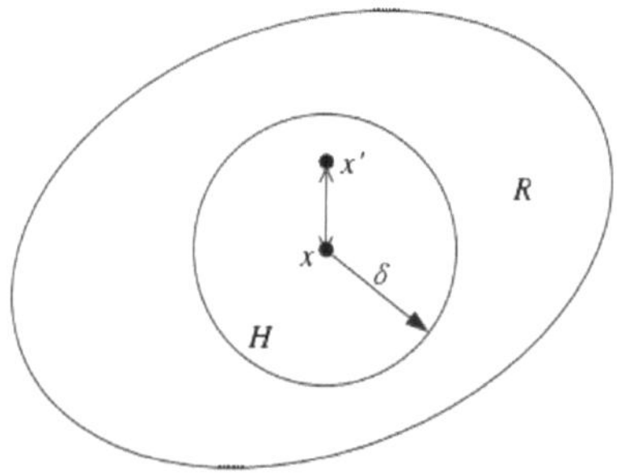


**Figure 1.** The peridynamic horizon of a material point ($H$), and the whole computing domain ($R$).

Each material point only interacts with other material points in its neighborhood. When a point exceeds the neighborhood range $\delta$, then $f(x',x,t) = 0$. Silling proposed the following expression for the force function in elastic isotropic materials [14]:

$$f(x',x,t) = \frac{\xi + \eta}{|\xi + \eta|}cs \tag{2}$$

where $\eta = u' - u$ is the relative displacement vector, $\xi = x' - x$ is the relative position vector and $c$ represents the micro-elastic modulus of the bond. Moreover, $s$ denotes the elongation of the bond, which is similar to the strain in conventional mechanics. It should be indicated that the parameter $c$ can be obtained through the strain energy density of PD form equal to the strain energy density of the classical elastic theory. The parameter $c$ of an isotropic material subjected to the plane stress is $c = 9E/(\pi h\delta^3)$, where $E$ is the Young's modulus, $\delta$ is the size of the neighborhood and $h$ denotes the distance between two material points.

### 2.2. Definition of Bond Break and Damage

In the bond-based peridynamic basic theory, the material failure is judged by the bond break. Currently, two judgment methods have been proposed for the bond break between material points. The first method is to judge based on the critical elongation $s_0$, and the second way is to judge based on the critical energy density. In the present study, the critical elongation is selected to judge the failure. To this end, the relative elongation is obtained when the bond between the material points is subjected to force. When the relative elongation exceeds the critical value $s_0$, the bond between the material points breaks and the specimen fails. Under this circumstance, there is no longer any interaction between points. The stress of the critical elongation in the plane can be calculated through the following expression:

$$s_0 = \sqrt{\frac{4\pi G_0}{9E\delta}} \tag{3}$$

where $G_0$ is the energy release rate during the crack propagation and $E$ is the Young's modulus. The correlation between the fracture toughness $K_{IC}$, the critical energy release rate $G_0$ and the Young's modulus of the material is as follows:

$$G_0 = \frac{{K_{IC}}^2}{E} \tag{4}$$

The local damage of the material point can be judged by the following damage coefficient:

$$\mu(x', x, t) = \begin{cases} 1 & \text{if } s \le s_0 \\ 0 & \text{if } s > s_0 \end{cases} \tag{5}$$

$$\varphi(x,t) = 1 - \frac{\int\limits_{H_x} \mu(x', x, t) dV_{x'}}{\int\limits_{H_x} dV_{x'}} \tag{6}$$

when $\varphi = 1$, the bond between the reference material point and other material points in the domain has been broken. On the contrary, $\varphi = 0$ indicates that no bond has been broken. Accordingly, the range of the damage coefficient is $[0,1]$.

### 2.3. Improved Bond-Based Peridynamics Model for FGM Analysis

An FGM is a heterogeneous material with variable spatial material composition. Consequently, material properties change along the spatial coordinates. Accordingly, the material properties of an FGM should be considered as spatial functions such as $\rho(x,y)$, $E(x,y)$ and $K_{IC}(x,y)$. Therefore, the material properties can be expressed in the form below:

$$\beta(x,y) = \beta(\beta_0, x, y) \tag{7}$$

where $\beta$ can be $\rho$, $E$ or $K_{IC}$. Similarly, $\beta_1$ can be replaced by $\rho_1$, $E_1$ or $K_{IC1}$. (The initial value of the function) In the PD method, the bond material properties between material points $x_i(x_i, y_i)$ and $x_j(x_j, y_j)$ should be obtained in these calculations. Generally, the average of material parameters for two material points can be used to calculate the corresponding PD bond material properties of the FGM. In this regard, the material properties of the bond model solved by the average value can be expressed as:

$$\beta(x_i, x_j) = 0.5(\beta(x_i, y_i) + \beta(x_j, y_j)) \tag{8}$$

The calculated average material parameters are then used to calculate the corresponding micro-modulus and critical elongation in the PD model:

$$c(x_i, x_j) = \frac{9E(x_i, x_j)}{\pi h \delta^3} \tag{9}$$

$$s_0(x_i, x_j) = \sqrt{\frac{4\pi G_0(x, \hat{x})}{9E\delta}} \tag{10}$$

However, the material parameters of FGMs do not change uniformly in space. Therefore, although the average value method is convenient to solve, it cannot accurately reflect the variations in the material parameters.

Figure 2 illustrates a simplified bond model, and shows that its material parameters, such as the Young's modulus, follow a function along the length direction $E(x)$.

**Figure 2.** Simplified diagram of bonds between material points.

When the two ends of the rod are under tension, the rod elongation can be calculated through the following expression:

$$\Delta l = \int_0^{l_0} \frac{P}{E(x)A} dx \tag{11}$$

where $l_0$ is the distance between two material points and $\Delta l$ is the amount of change in the distance between the two materials, which can be calculated as follows:

$$\Delta l = \frac{Pl_0}{E_0 A} \tag{12}$$

where $E_0$ is the equivalent Young's modulus of the rod. The equivalent Young's modulus can then be obtained by combining Equations (11) and (12):

$$E_0 = \frac{l_0}{\int_0^{l_0} 1/E(x)dx} \tag{13}$$

$E(x)$ is a function reflecting the variations in material parameters. It is an enormous challenge to solve this integral expression theoretically. Therefore, a numerical integration method can be used to solve it. Subsequently, the equivalent parameters between the two material points can be obtained and the corresponding micro-modulus and critical elongation can be solved.

## 3. Model Checking

This section intends to verify the accuracy of the improved PD model established in Section 2. To this end, the crack growth in an organic glass plate and an FGM beam under different loads are simulated.

### 3.1. Simulating the Crack Propagation in a Plexiglass Sheet

#### 3.1.1. Program Verification

In order to verify the accuracy of the near-field dynamics program, the crack propagation problem in a rectangular plexiglass sheet with prefabricated cracks under a uniformly distributed load at both ends is simulated. The board is 250 mm long and 100 mm wide, and the position of the pre-crack is shown in Figure 3.

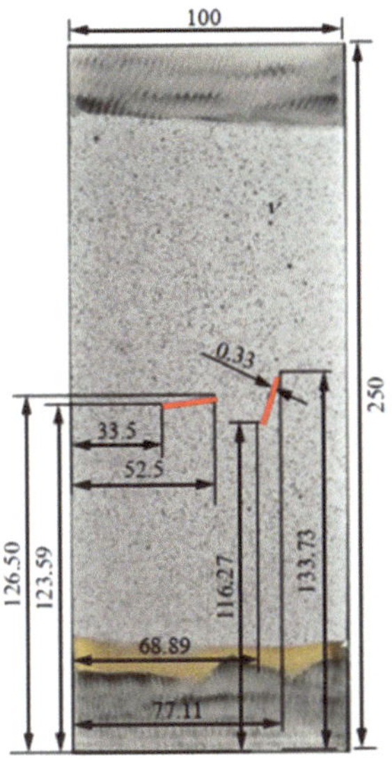


**Figure 3.** Schematic diagram of the pre-crack position [15].

The red solid line in Figure 3 shows the position of the prefabricated crack. Parameters of the plexiglass are presented in Table 1.

**Table 1.** Material parameters of plexiglass.

| Material Parameter | Plexiglass |
|---|---|
| Elastic modulus E (GPa) | 2.633 |
| Density $\rho$ ($kg/m^3$) | 1200 |
| Fracture toughness $K_{IC}$ ($MPa{\cdot}m^{1/2}$) | 2.2 |

The upper and lower ends of the plate are subjected to a uniform load of 13 MPa, and the particle spacing is 0.0005 m. Moreover, the neighborhood size is set to three times the particle spacing, and the calculation time step is set to 0.1 microseconds. The influence function is defined as follows [16]:

$$g(\lambda, \Delta) = \left(1 - (\lambda/3\Delta)^2\right)^2 \tag{14}$$

where $\lambda$ is the distance between two particles at each moment, and $\Delta$ is the particle spacing set in the initial model. Equation (14) is then multiplied by the corresponding force function to get the corresponding particle interaction. Figure 4a,b shows the experimental and calculated final forms of the cracks, respectively.

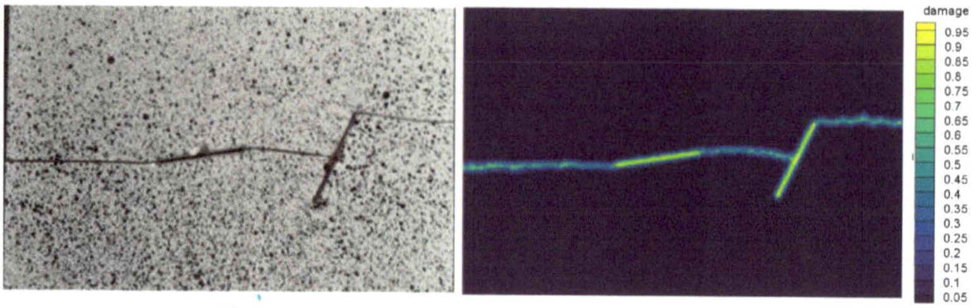


(a) (b)

**Figure 4.** Crack path at complete failure obtained from the (**a**) experiment [16] and (**b**) numerical simulation.

Comparison of the results from the experiment and numerical simulation reveals that there is a good consistency, indicating that the established method is accurate.

Next, the parameters affecting the accuracy of the PD calculation, including the influence function, neighborhood range and particle distance, are investigated. The main purposes of this investigation are to apply (1) the influence function, (2) different neighborhood ranges and (3) different particle spacing on the crack path of the plexiglass plate. The calculation results are then compared with the experimental results to evaluate the influence of the different parameters on the simulation accuracy and calculation efficiency.

#### 3.1.2. The Effect of the Influence Function on the Calculation Accuracy

Within the neighborhood of a particle, the distance between the reference particle and other particles is different, resulting in different interactions between particles. Generally, the longer the distance, the smaller the interaction. In the present study, only the influence function shown in Equation (14) is applied to calculate the crack propagation path of the plexiglass plate with and without the influence function. The obtained results in this regard are shown in Figure 5. Except for the influence function, other material properties and dimensions of the calculation model are the same as those in Section 3.1.1.

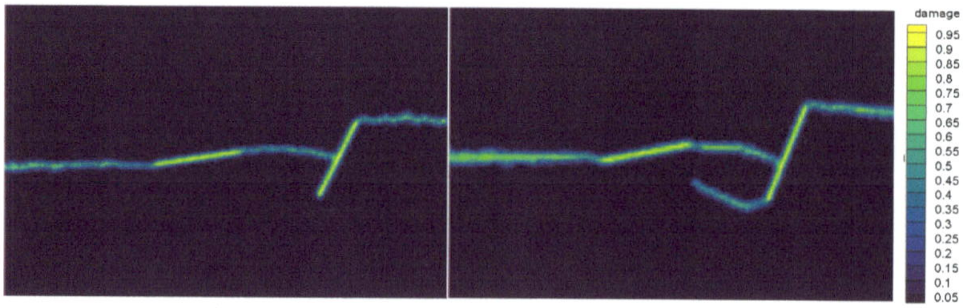


(a) (b)

**Figure 5.** Crack path diagram (**a**) with the influence function, and (**b**) without the influence function.

Figure 5a,b reveals that the calculated results after setting the influence function are in better agreement with the experiment shown in Figure 4. Accordingly, it is concluded that setting an appropriate influence function can improve the calculation accuracy.

#### 3.1.3. The Influence of the Neighborhood Range on Calculation Accuracy

In the numerical calculation of the bond-based PD theory, it is necessary to set the size of the neighborhood of the material points to determine the distance at which a material point interacts with other points. For example, when the neighborhood range is set as particle spacing $\delta = \Delta d$, this means that each particle only interacts with the nearest particle nearby. This issue is shown in the following two-dimensional case (Figure 6).

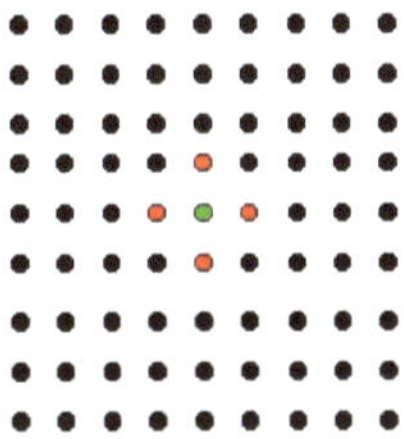

**Figure 6.** The peridynamic horizon of the single-particle spacing.

In this section, the crack paths are calculated in different neighborhoods, including $\delta = 1\Delta d$, $\delta = \sqrt{2}\Delta d$, $\delta = 2\Delta d$, $\delta = \sqrt{5}\Delta d$, $\delta = 2\sqrt{2}\Delta d$, $\delta = 3\Delta d$, $\delta = \sqrt{10}\Delta d$, $\delta = \sqrt{13}\Delta d$. It should be indicated that in all calculations, the same model as that in Section 3.1.1 is applied.

Figure 7 shows that the best calculation results can be obtained when the neighborhood range is three times the particle spacing or more. However, as the neighborhood range increases, the corresponding calculation time increases significantly. Accordingly, the neighborhood range three times the particle spacing is considered as the best spacing with the best accuracy and efficiency.

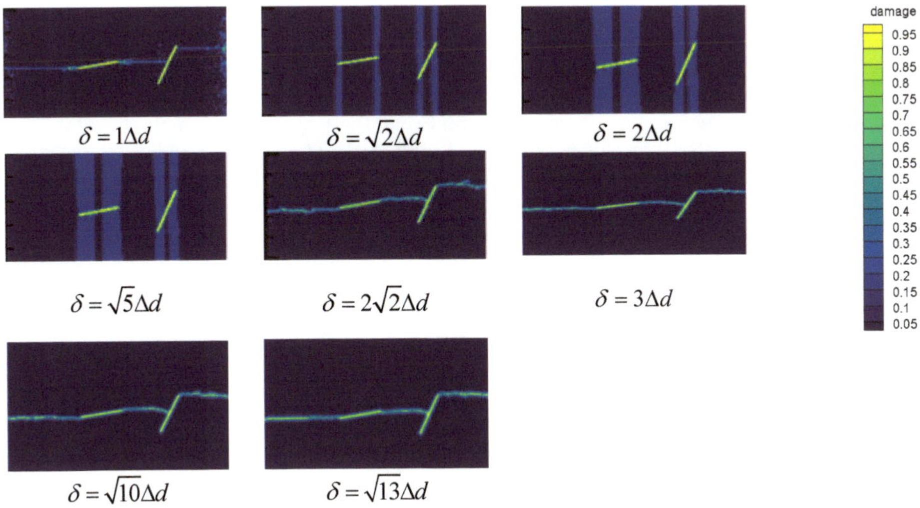


**Figure 7.** Crack path with different neighborhoods.

### 3.1.4. The Effect of the Particle Spacing on the Calculation Accuracy

Studies show that particle spacing greatly affects the numerical calculation results of the PD theory. Generally, the smaller the grid size, the higher the accuracy of the calculation result. However, the grid size cannot be reduced indefinitely because this will greatly increase the calculation time. Therefore, it is necessary to set an appropriate particle spacing in the calculation in order to meet the accuracy and efficiency requirements of the calculation. In this regard, the crack propagation path of the plexiglass plate with different particle spacing, including $\Delta d = 2 \times 10^{-3}$ m, $\Delta d = 1 \times 10^{-3}$ m and $\Delta d = 5 \times 10^{-4}$ m, are simulated in this section. Note that except for the particle spacing, all parameters are the same as those in Section 3.1.1.

Figure 8 indicates that for the particle spacing of $\Delta d = 2 \times 10^{-3}$ m and $\Delta d = 1 \times 10^{-3}$ m, the crack path differs from the experiment. On the other hand, when the particle spacing is set to $5 \times 10^{-4}$ m, a good agreement can be achieved with the experiment. Further reduction in the grid size greatly increases the calculation time, thereby adversely affecting the calculation efficiency. Therefore, it is concluded that when the grid size is less than or equal to the order of $10^{-4}$ m, the accuracy requirements of the calculation can be met.

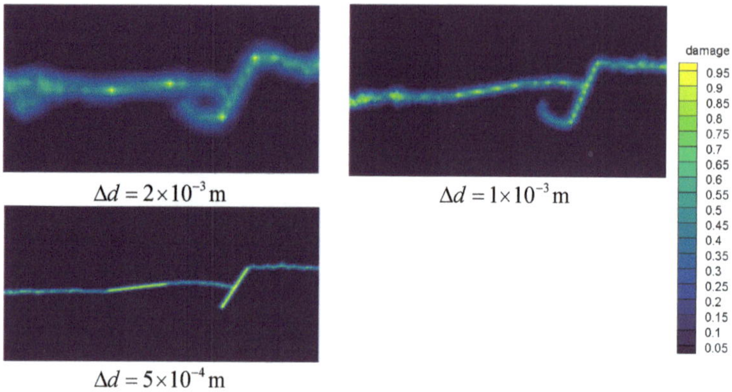


**Figure 8.** Crack path diagrams with different particle spacing.

## 3.2. *Simulation of Crack Propagation in FGM Beam under a Four-Point Bending Load*

In order to further verify the validity and reliability of the improved PD model, a four-point bending pre-cracked FGM specimen is simulated. The section size of the test specimen is 120 mm × 22 mm. Under the action of a four-point bending load, there is a material transition zone with a length of W = 37 mm in the middle of the specimen. Figure 9 schematically shows the model. The left side of the transition area is pure epoxy, while the right side is pure glass. The initial length of the crack is a = 5.5 mm, b = 14 mm. The particle distance and the calculation time step are set to 0.0025 m and 0.02 ms, respectively. Moreover, the neighborhood radius is three times the particle distance and the load P is 100 kN.

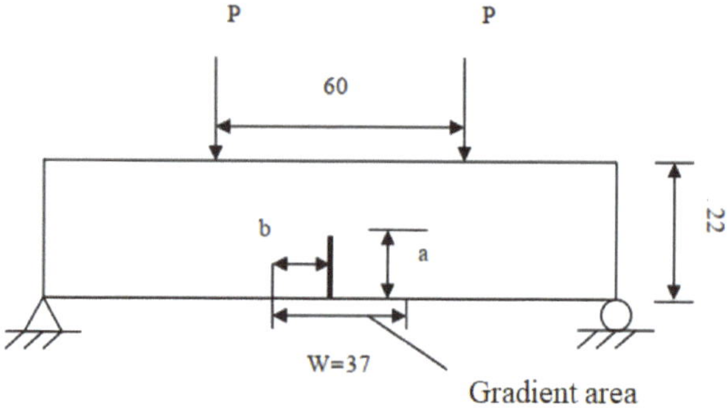


**Figure 9.** Schematic diagram of the model.

The material parameters of the studied FGM are presented in Table 2.

**Table 2.** Material parameter table of the FGM.

| Material Parameter | Epoxy Resin | Glass |
|---|---|---|
| Elastic modulus $E$ (GPa) | 3 | 8.6 |
| Density $\rho$ (kg/m$^3$) | 1200 | 1850 |
| Fracture toughness $K_{IC}$ (MPa·m$^{1/2}$) | 1.2 | 2.6 |

The material parameters of the gradient region change in the form of a power function as below:

$$\beta(x,y) = \beta_0 + (\beta_w - \beta_0)\left(\frac{x}{W}\right)^4 \quad (15)$$

Comparison of Figure 10a–d indicates that the calculation results obtained from the proposed model are in good agreement with the experiment.

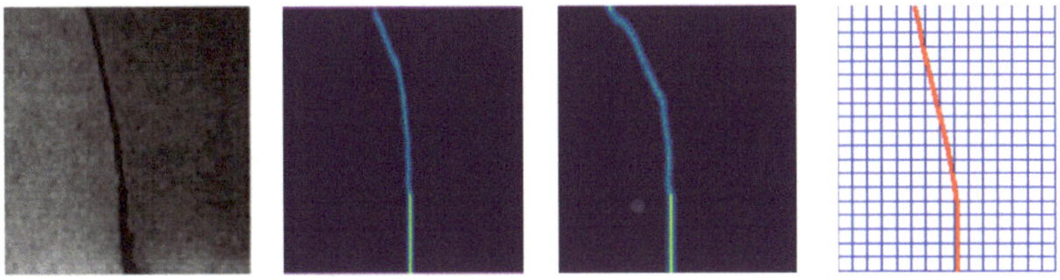

**Figure 10.** Crack path diagram: (**a**) Experimental result [17], (**b**) calculation results obtained from the improved PD model, (**c**) calculation results from the standard PD model and (**d**) extended finite element calculation results.

In order to compare the above crack path more clearly and intuitively, the corresponding data are extracted and a chart is prepared accordingly.

Figure 11 shows that in the early stages of crack propagation, the calculation model in this paper does not reflect the corresponding superiority. However, as the crack continues to grow, it can be observed that the calculation results of the traditional PD model gradually deviate from the experimental results. When the complete specimen is fractured, with the experimental results as the reference point (breaking point at top), the calculation error of the traditional PD model is 16.1% whereas the error of the model in this paper is 4.3%. Therefore, it can be considered that the model proposed in this paper improves the calculation accuracy to a certain extent.

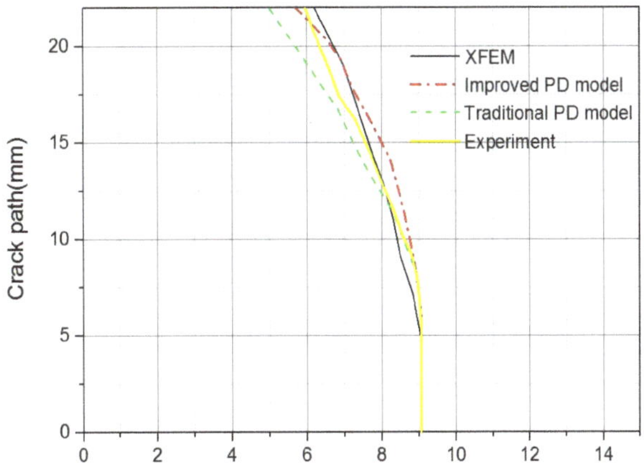


**Figure 11.** Crack path diagram.

## 4. Conclusions

In the present study, the integral equivalent solution method is applied to propose an improved bond-based PD model and calculate the material parameters of the bond. Compared with the conventional average solution method, the proposed method is more in line with actual material property gradient changes. The proposed model is applied to simulate the dynamic crack growth of ae plexiglass plate and an FGM beam under different loads. The calculation results are then compared with the experiment, and the model accuracy is evaluated. The influence of different PD parameters on the simulation accuracy was discussed, and it is concluded that the best results can be achieved when the affecting parameters are set as follows: 1. the neighborhood range is set to three times the particle spacing, 2. an appropriate influence function should be considered in the calculations and 3. the distance between particles is on the order of $10^{-4}$ m or less.

It is found that compared with conventional methods, the results of the proposed model are more in line with the experiment.

**Author Contributions:** Conceptualization, L.M.; data curation, L.M.; formal analysis, L.M.; funding acquisition, Q.L. and L.L.; methodology, L.M., Q.L. and L.L.; software, L.M.; writing—original draft, L.M.; writing—review and editing, L.M. and Q.L. All authors have read and agreed to the published version of the manuscript.

**Funding:** This research was funded by the National Natural Science Foundation of China (Grant Nos. 11802214 and 11972267) and the Fundamental Research Funds for the Central Universities (WUT: 2018IB006 and WUT: 2019IVB042).

**Institutional Review Board Statement:** Not applicable.

**Informed Consent Statement:** Not applicable.

**Acknowledgments:** We would like to thank Zhenghong Huang and Rousseau C.E for their experiments. This article has obtained their permission to use their experimental results Their experimental results verified the correctness of the calculated results in this paper. Their experiments provide important support for this paper.

**Conflicts of Interest:** The authors declare no conflict of interest. The funders had no role in the design of the conflicts of interest: study; in the collection, analysis, or interpretation of data; in the writing of the manuscript; or in the decision to publish the results.

## References

1. Atkinson, C. Discussion: "the crack problem for bonded nonhomogeneous materials under antiplane shear loading" (erdogan, f., 1985, asme j. Appl. Mech., 52, pp. 823–828). *J. Appl. Mech.* **1986**, *3*, 729. [CrossRef]
2. Delale, F.; Erdogan, F. The crack problem for a nonhomogeneous plane. *J. Appl. Mech.* **1983**, *3*, 609–614. [CrossRef]
3. Jin, Z.; Noda, N. Crack-tip singular fields in nonhomogeneous materials. *J. Appl. Mech.* **1994**, *3*, 738–740. [CrossRef]
4. Marur, P.R.; Tippur, H.V. Numerical analysis of crack-tip fields in functionally graded materials with a crack normal to the elastic gradient. *Int. J. Solids Struct.* **2000**, *38*, 5353–5370. [CrossRef]
5. Kawasaki, A.; Watanabe, R. Thermal fracture behavior of metal/ceramic functionally graded materials. *Eng. Fract. Mech.* **2002**, *14–16*, 1713–1728. [CrossRef]
6. Nishioka, T.; Tokudome, H.; Kinoshita, M. Dynamic fracture-path prediction in impact fracture phenomena using moving finite element method based on delaunay automatic mesh generation. *Int. J. Solids Struct.* **2001**, *30*, 5273–5301. [CrossRef]
7. Fabbrocino, F.; Funari, M.F.; Greco, F.; Lonetti, P.; Luciano, R.; Penna, R. Dynamic crack growth based on moving mesh method. *Compos. Part B Eng.* **2019**, *174*, 107053. [CrossRef]
8. Dirik, H.; Yalinkaya, T. Crack path and life prediction under mixed mode cyclic variable amplitude loading through XFEM. *Int. J. Fatigue* **2018**, *114*, S440411598. [CrossRef]
9. Silling, S.A. Reformulation of elasticity theory for discontinuities and long-range forces. *J. Mech. Phys. Solids* **2000**, *48*, 175–209. [CrossRef]
10. Ha, Y.D.; Bobaru, F. Studies of dynamic crack propagation and crack branching with peridynamics. *Int. J. Fract.* **2010**, *1*, 229–244. [CrossRef]
11. Ha, Y.D.; Lee, J.; Hong, J.W. Fracturing patterns of rock-like materials in compression captured with peridynamics. *Eng. Fract. Mech.* **2015**, *144*, 176–193. [CrossRef]
12. Cheng, Z.; Zhang, G.; Wang, Y.; Bobaru, F. A peridynamic model for dynamic fracture in functionally graded materials—sciencedirect. *Compos. Struct.* **2015**, *133*, 529–546. [CrossRef]
13. Tan, Y.; Liu, Q.; Zhang, L.; Liu, L.; Lai, X. Peridynamics model with surface correction near insulated cracks for transient heat conduction in functionally graded materials. *Materials* **2020**, *6*, 1340. [CrossRef] [PubMed]
14. Silling, S.A.; Askari, E. A meshfree method based on the peridynamic model of solid mechanics. *Comput. Struct.* **2005**, *83*, 1526–1535. [CrossRef]
15. Zhenghong, H.; Shouchun, D.; Haibo, L.; Chong, Y. Experimental study and numerical simulation of crack propagation process and intersection mode under tensile load. *Chin. J. Rock Mech. Eng.* **2019**, *S1*, 2712–2723.

16. Dan, H.; Lu, G.; Wang, C.; Qiao, P. An extended peridynamic approach for deformation and fracture analysis. *Eng. Fract. Mech.* **2015**, *141*, 196–211.
17. Rousseau, C.E.; Tippur, H.V. Compositionally graded materials with cracks normal to the elastic gradient. *Acta Mater.* **2000**, *16*, 4021–4033. [CrossRef]

materials

Article

# Failure Analysis and Prevention of Extraction Column for Methyl Methacrylate Production: Application of the 'Safety Design' Concept

**Sheng-Hui Wang [1,2], Yi Gong [1], Qi Tong [3], Xiao-Lei Yang [1], Jia-Hao Shen [3] and Zhen-Guo Yang [1,*]**

1 Department of Materials Science, Fudan University, Shanghai 200433, China; wangwish6@163.com (S.-H.W.); gongyi@fudan.edu.cn (Y.G.); 18110300022@fudan.edu.cn (X.-L.Y.)
2 Shanghai Institute of Special Equipment Inspection and Technical Research, Shanghai 200062, China
3 Department of Aeronautics and Astronautics, Fudan University, Shanghai 200433, China; tongqi@fudan.edu.cn (Q.T.); 18210290006@fudan.edu.cn (J.-H.S.)
* Correspondence: zgyang@fudan.edu.cn

**Abstract:** To ensure safety and prevent failure of engineering equipment throughout its lifespan, the concept of 'Safety Design' is proposed, which covers all the cradle-to-grave phases of engineering equipment, considers at least ten essential factors of failure causes, and conducts root cause analysis at three different scales, in order to proactively control the safety risks before the occurrence of failure rather than passively conduct the remedial measures after failure. Herein, in order to demonstrate how to implement this effective and efficient concept in engineering practice, a case study of failure analysis and prevention is addressed on the extraction column in the production line for methyl methacrylate. Based on the analysis results, the causes were finally determined to be all derived from the stages before operation, including inappropriate design, limited quality inspection of fabrication and installation. Pertinent countermeasures were then proposed from the 'Safety Design' point of view, which would not only solve the failure problem for this sole equipment but also contribute to safety risk control of other engineering equipment before operation.

**Keywords:** safety design; failure analysis; extraction column; methyl methacrylate; finite element analysis

**Citation:** Wang, S.-H.; Gong, Y.; Tong, Q.; Yang, X.-L.; Shen, J.-H.; Yang, Z.-G. Failure Analysis and Prevention of Extraction Column for Methyl Methacrylate Production: Application of the 'Safety Design' Concept. *Materials* **2021**, *14*, 4234. https://doi.org/10.3390/ma14154234

Academic Editor: Jason Anthony Scott

Received: 12 May 2021
Accepted: 27 July 2021
Published: 29 July 2021



## 1. Introduction

To ensure the safety of engineering equipment, the philosophy of full life cycle (FLC) or the methodology of life cycle management (LCM) is universally accepted and adopted in industry [1–3], which covers all the cradle-to-grave phases of equipment, including design, manufacturing, inspection, storage, transportation, installation, commissioning, operation, maintenance, overhaul, shutdown, and decommissioning [4], and manages both the physical factors (degradation, malfunction and failure) and the non-physical factors (asset, economy and staff). Thereinto, design is the foremost phase, which mainly involves siting, configuration, layout, materials, loads and environments, etc., and accounts for 79% of the failure causes of engineering equipment accidents [5]. Operation is the pivotal phase where engineering equipment fulfills its functionalities as designed and is intimately related to materials performance, management strategy and personnel skills. Nevertheless, during the operation phase, engineering equipment is usually obliged to be shutdown temporarily or even long-term if unexpected and/or premature failures occur due to the improper actions taken before this phase, e.g., stress corrosion cracking and crevice corrosion from improper design [6], fracture from improper manufacturing [7], localized erosion from improper installation [8], and atmospheric corrosion from improper storage [9], etc., based on our erstwhile experience.

Failure analysis is the action commonly undertaken to identify the causes of failed engineering equipment and propose the countermeasures for prevention of the same or similar cases from reoccurrence [10–12]. Given that failure greatly influences the reliability, safety, economy and longevity of engineering equipment, it is optimal to proactively control and even eliminate the safety risks before failure [13], rather than to passively implement the remedial measures after failure. This concept can be regarded as the underlying objective of failure analysis and well conforms to the strategy of the FLC philosophy and the LCM methodology in nature. On this basis, we put forward the 'Safety Design' concept in brief, which integrates failure analysis and prevention into three aspects. (1) Stage. For convenience, we classify all the cradle-to-grave phases of engineering equipment into three basic stages, including the upstream stage, the midstream stage and the downstream stage, which respectively concentrates on design safety, fabrication safety and service safety, as presented in Figure 1. (2) Scope. To correctly identify the causes of failed engineering equipment, we summarize at least ten essential factors that are necessitated to be taken into account in the failure analysis process, including design, material, manufacture, installation, inspection, operation, maintenance, environment, transportation and storage, and thereby establish the fishbone (Ishikawa) diagram, as displayed in Figure 2. (3) Scale. For the sake of effectiveness, efficiency and economy, in the failure analysis process, we conduct observation of morphologies of failed engineering equipment at three different scales in sequence, including a preliminary analysis in the macroscopic scale (>$10^{-2}$ m) to determine the failure modes (e.g., corrosion, wear, fracture and distortion), a detailed analysis in the mesoscopic scale ($10^{-5}$~$10^{-2}$ m) to confirm the failure forms (e.g., pits, dents, cracks, striations), and an in-depth analysis in the microscopic scale (<$10^{-5}$ m) [14] to differentiate the failure mechanisms (e.g., pitting corrosion, abrasive wear, stress corrosion cracking, fatigue). Once the failure causes are identified at the upper scales, no further analysis is necessary at the subsequent lower scales.

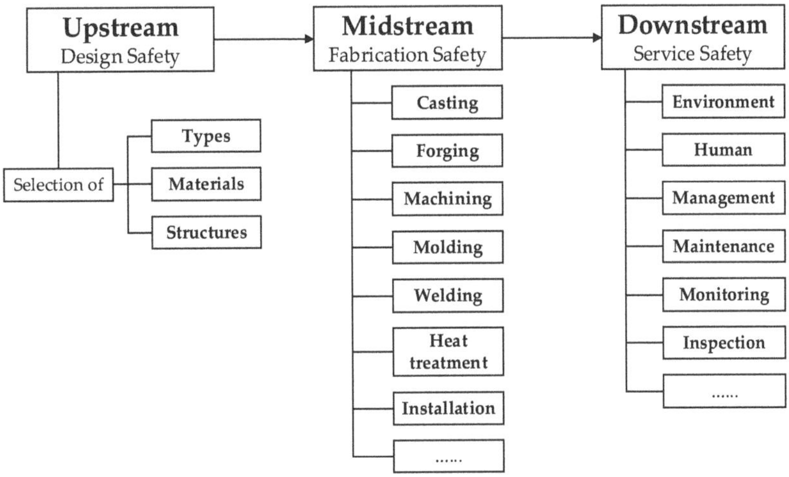


**Figure 1.** The 'Safety Design' concept in the three stages of engineering equipment.

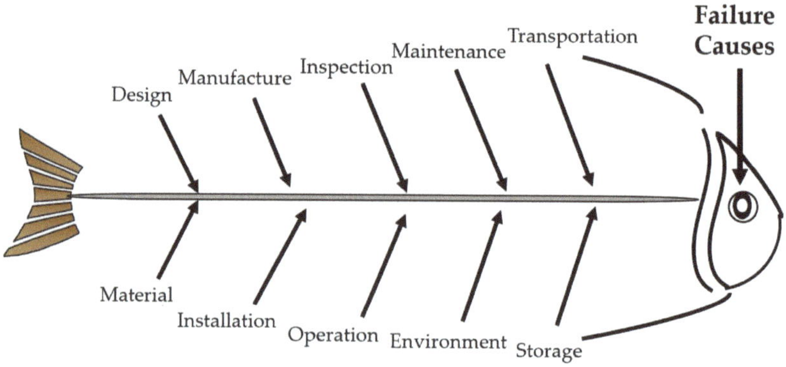


**Figure 2.** The fishbone diagram of failure analysis for engineering equipment.

To demonstrate how to put this 'Safety Design' concept into practice for engineering equipment, this paper presents a case study of failure analysis and prevention on one piece of petrochemical equipment, the extraction column in the production system of methyl methacrylate (MMA). According to the design requirement, the projected lifetime of it was at least twenty years. However, unexpected abnormal wear was discovered on the inner wall of its shell during the routine overhaul after serving for only four years. Since the normal operation of the extraction column had been affected, which would probably even threaten the safety of the whole system, it was therefore shut down at once and the standby one was put into service. At the same time, in order to prevent the reoccurrence of the same case, failure analysis was urgently required as well. By referring to our previous experiences on cause analysis of failed engineering equipment in the last several years, comprehensive investigation and systematic analysis was immediately carried out. As a result, in addition to the cause identification, the countermeasures focusing on the phases before operation were also proposed for failure prevention, including design modification, manufacturing improvement and installation optimization. This paper is a sequel to our previous cause analysis of localized abnormal wear for this same piece of equipment [15], but it can be taken as an example to demonstrate the 'Safety Design' concept in engineering practice. The achievements of this paper would not only enrich the database of failure analysis cases [16–18] but also have reference value for scholars in relevant fields to utilize this 'Safety Design' concept in the failure analysis and prevention of engineering equipment.

## 2. Experimental

Herein, one certain petrochemical company adopts the C4 direct oxidation method to produce MMA with annual output of over 100,000 tons. This method consists of three main steps in sequence, as listed in Figure 3. It initiates from the oxidation of the C4 feedstock isobutene ($C_4H_8$), as the name implies; progresses with the purification of the intermediate product methylacrylic acid (MAA); and terminates at the esterification of MAA for producing the end product MMA. The extraction column is located in the second step, and its main functionality is to purify the crude MAA by mixing the MAA aqueous solution with the extraction solvent heptane ($C_8H_{18}$). As exhibited in Figure 4a, it is in the shape of a vertical cylindrical shell with the size of 21,300 × 2000 × 10 mm$^3$ (height × diameter × thickness), and the matrix material is 316 stainless steel. Inside it, an active tray is installed, as shown in Figure 4b,c, which is controlled by the main shaft in the center of the extraction column to achieve a purification function through vertical reciprocating movement with the liquids under the designed frequency of 90 per minute. Furthermore, it is displayed in Figure 5a that the tray is further divided into four modules along the vertical direction. Each module contains 36 sieve plates, and the plates are separated into groups of 6 by a baffle made of polytetrafluoroethylene (PTFE) for protection of the tray from direct collision onto the shell, as shown in Figure 5b. According to the design drawing, the diameter of the PTFE baffle is 1985 mm, that is, a 7.5 mm-wide clearance, (2000−1985)/2, is reserved between the PTFE baffles and the inner wall of the shell.

After serving for four years, it was detected that the operational frequency of the tray gradually decreased. As a matter of fact, the phenomenon of frequency decrease had already been anticipated in the design because the products and byproducts would inevitably deposit and foul on the sieve plates of the column, and consequently, the actual operational frequency of the tray was set and maintained in the range of 60 to 70 per minute. However, this time the frequency decreased by half, falling into the range of 30 to 40 per minute, and could not be increased any more. Undoubtedly, the extraction efficiency of the column and thereby the production capacity of the system would be affected if this problem was not adequately solved. Hence, the extraction column was shut down at once, and the standby one was put into service. In addition, an inspection was conducted by the field personnel on the interior of the failed column to identify the causes. It was reported by them that some of the PTFE baffles, even the sieve plates, were directly contacting with the shell of the column, and lots of strip-shaped traces were left on the inner wall of the shell. It is explicit that the direct contact between the baffles/sieve plates and the shell must be the reason for the frequency decrease of the tray, and the problem was thus accordingly shifted to identify the causes and consequences. To this end, the whole tray was pulled out from the shell for further investigation (Figure 6a). Thereby, lots of parallel wear traces indeed emerged on the inner wall of the shell, as displayed in Figure 6b.

In normal conditions, it is impossible for the tray to contact the inner wall of the shell on account of the 7.5 mm-wide clearance. Hence, the first idea of the field personnel was to ascribe the wear to the inclination of the shell. If so, the whole extraction column must be completely refurbished or even replaced and would induce unaffordable economic losses to the company. To ascertain/eliminate that cause, failure analysis was then carried out by us.

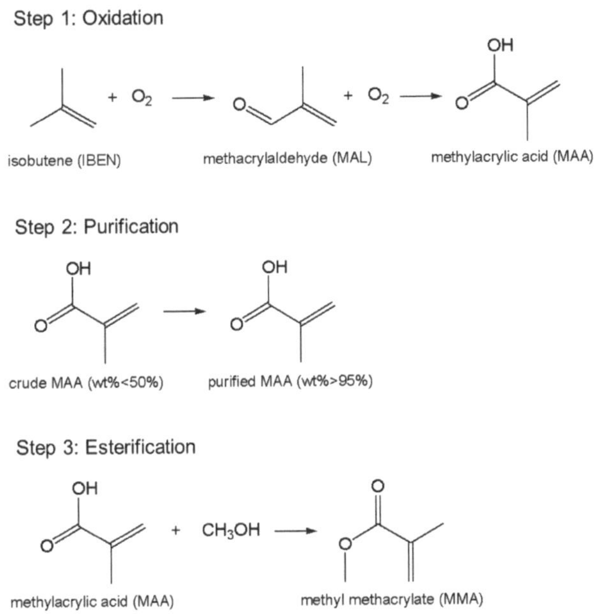


**Figure 3.** The main steps of C4 direct oxidation method to produce MMA.

(**a**) in the vertical cylindrical shell shape (**b**) the tray pulled out from the shell (**c**) scheme of the tray and column

**Figure 4.** External appearances of the extraction column.

(**a**) four modules (**b**) sieve plates and PTFE baffle

**Figure 5.** External appearances of the tray.

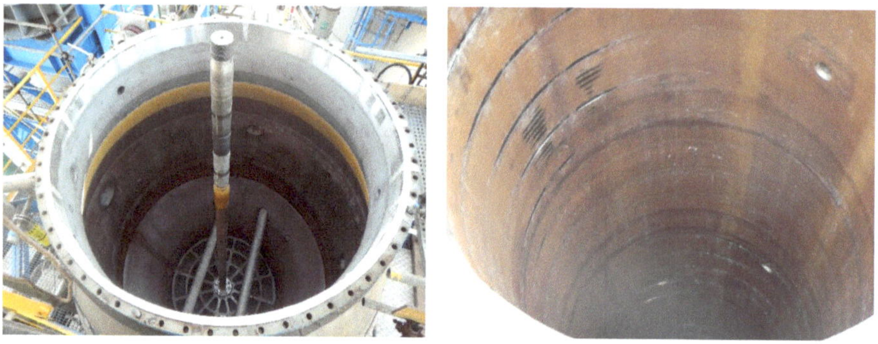

(**a**) the main shaft left (**b**) lots of parallel wear traces

**Figure 6.** External appearances of the shell inner wall after pulling out the tray.

Since this extraction column was an intact pressure vessel, it was not feasible to cut material samples from any component of it, especially the shell. Consequently, according to the 'Safety Design' concept, investigation could only be carried out at the macroscopic scale. Based on the analysis results, the safety risks in the downstream service stage were excluded. In other words, the failure causes were primarily derived from the other two stages, i.e., the upstream design stage and the midstream fabrication stage. Thus, to demonstrate the way to practice the 'Safety Design' concept, the following paragraphs will address the failure causes identification of this extraction column and propose countermeasures for failure prevention of similar engineering equipment in the future.

## 3. Results and Discussion

### *3.1. Failure Analysis*

The first step was to investigate the possibility of inclination of the shell of the extraction column. Thus, visual inspection was conducted on the inner wall of the shell. As displayed in Figure 7, the wear traces were discovered distributing on all the four main directions of 0°, 90°, 180° and 270° along the circumference, and on all the heights equal to the four modules of the tray. That is, the degree of wear between the different directions and heights showed no obvious difference; only in some areas, the trace depth reached as deep as 2.5 mm (Figure 7b). This phenomenon was solid evidence of excluding the possibility of inclination of the column shell, because it would have caused wear traces in one direction, and the wear extent would have gradually decreased downwards along the height.

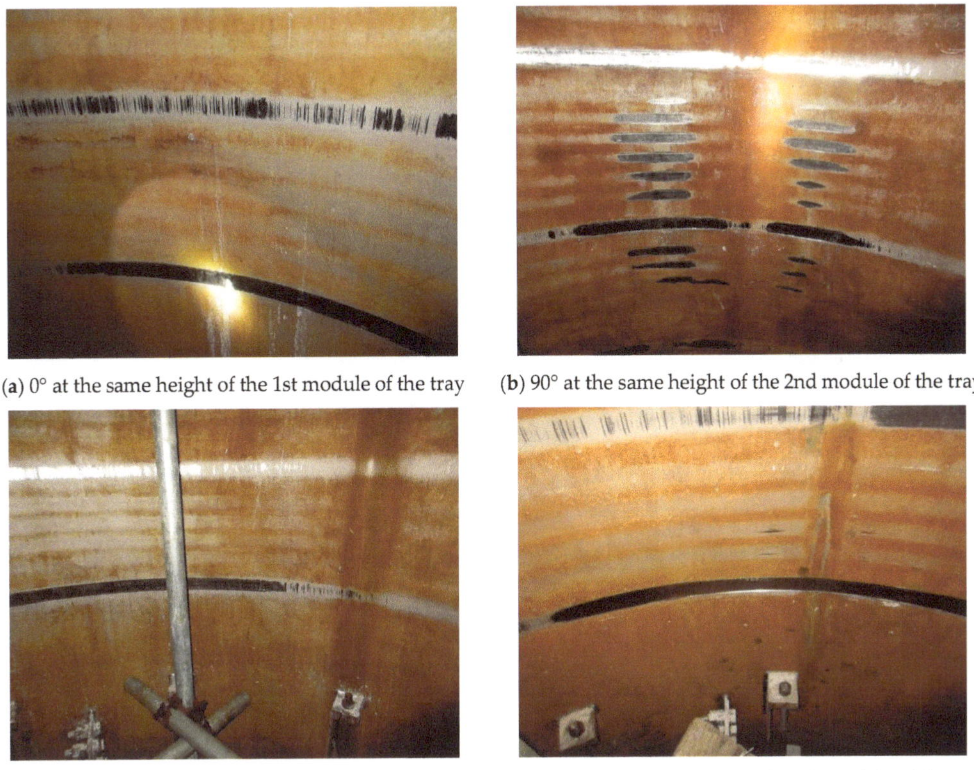

(**a**) 0° at the same height of the 1st module of the tray (**b**) 90° at the same height of the 2nd module of the tray

(**c**) 180° at the same height of the 3rd module of the tray (**d**) 270° at the same height of the 4th module of the tray

**Figure 7.** Wear traces on the inner wall of the shell.

While for the tray, wear phenomenon was also discovered. Compared with the intact PTFE baffles that had not suffered wear (Figure 8a), the ones that had suffered wear exhibited diverse morphologies such as adhesive wear (Figure 8b), curled strip (Figure 8c) and material loss (Figure 8d), etc., indicating quite severe wear.

Thus far, it can be basically ascertained that severe wear had indeed occurred between the tray (i.e., the sieve plates and the PTFE baffles) and the inner wall of the column shell, but the possibility of inclination of the shell was excluded. In other words, the causes should be revealed from other aspects.

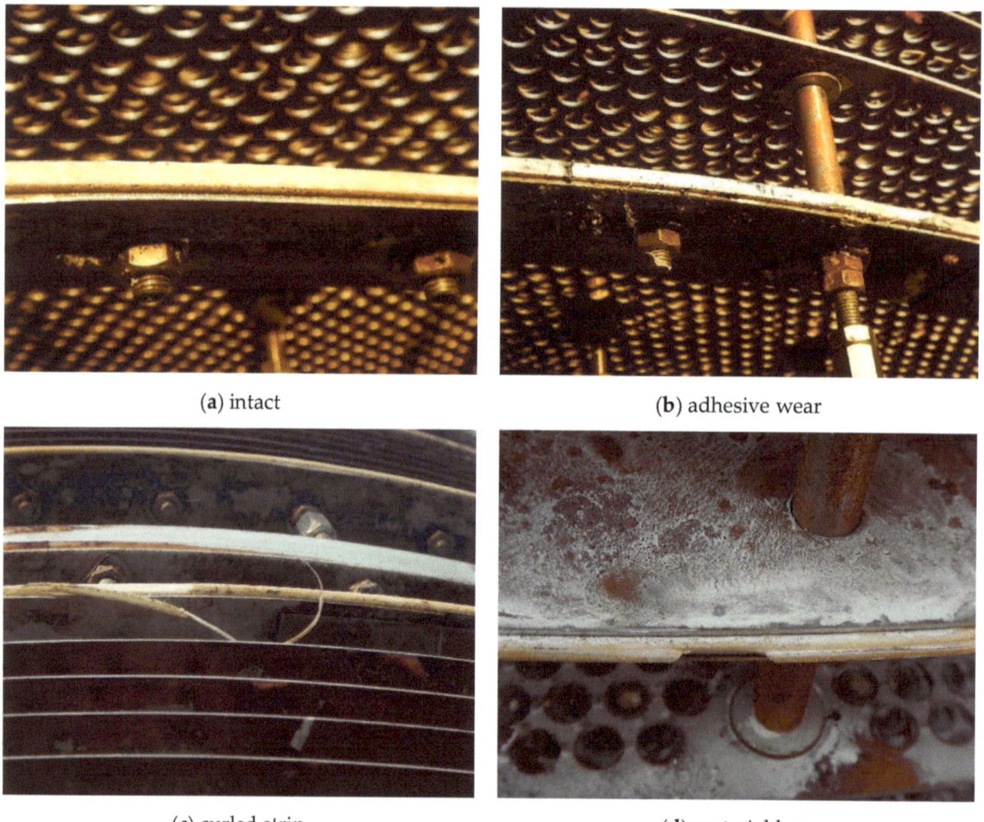

(**a**) intact (**b**) adhesive wear (**c**) curled strip (**d**) material loss

**Figure 8.** Diverse wear morphologies on the PTFE baffles.

### 3.2. Design Safety Analysis

As mentioned above, on the main shaft (as shown in Figure 6a) there is a tray equipped for extraction, which contains a total of 5 PTFE baffles and 144 sieve plates. Per the design drawing, the diameters of the baffles and the plates are 1985 and 1920 mm, respectively, while that of the main shaft is only 88.9 mm, less than 5% of the former two. Then, a problem naturally emerges of whether the main shaft has sufficient structural stiffness to hold the tray under the operational conditions. Hence, a finite element analysis was performed.

Since the weight of the tray has little effect on the main shaft, the load taken into consideration was mainly the impact from the two fluids, i.e., the MAA aqueous solution flowing downwards and the extraction solvent heptane flowing upwards. By using ABAQUS, the numerical model was established according to the design drawing, and only the main shaft with the tray involved in the vertical reciprocating movement was taken into consideration, as presented in Figure 9a. The impact of fluids was time-averaged and applied as static pressure on the surfaces of the sieve plates and the PTFE baffles, which can be deduced under the principle of mass and momentum conservation, as seen in Equation (1).

$$P = \frac{1}{2A_f}\left(1 - A_f^2\right)\left(2g\Delta h + A_f^2 v^2\right) \quad (1)$$

where $P$ is the momentum (kg·m/s), $A_f$ is the area ratio (dimensionless), $g$ is the gravitational acceleration (m/s$^2$), $\Delta h$ is the height difference (m) and $v$ is the velocity (m/s). All the components were made of 316 stainless steel except the PTFE gasket, and the materials parameters (Table 1) could be obtained from the materials handbook.

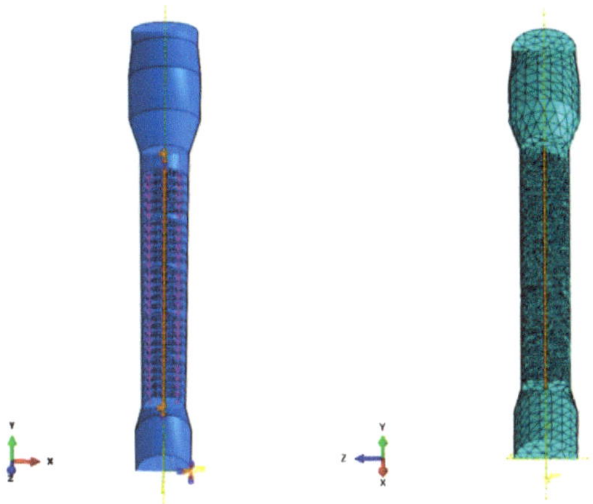

(**a**) loading conditions (**b**) after meshing

**Figure 9.** The finite element model of the extraction column.

**Table 1.** Physical and mechanical properties of 316 stainless steel.

| Properties | Density (kg/m$^3$) | Elastic Modulus (GPa) | Tensile Strength (MPa) | Yield Strength (MPa) | Elongation (%) |
|---|---|---|---|---|---|
| Values | 7980 | ~190 | ~580 | ~205 | ≥40 |

In the model, the shell was considered rigid and coarsely meshed with 7223 three-dimensional rigid triangular elements (R3D3 from Abaqus element library). To prevent the sieve plates and the PTFE baffles from yielding under the static pressure, they were modeled by stiff materials compared with the main shaft, but all these three components were finely meshed with 242,130 tetrahedral elements C3D10, seen in Figure 9b. It should be noted that it generally will save much computational cost to use structural elements instead of continuum elements. Nonetheless, the coupling of degrees of freedom between nodes of different types of structural elements is very complicated and may lead to unrealistic results. In this study, we consider the contact of the plates and baffles with the rigid shell and the deformable shaft. It makes the construction of the corresponding constraint equations of the nodal degrees of freedom quite complex under the Kirchhoff assumption of shells. This is the reason for choosing continuum elements C3D10 for the simulation. As for the boundary conditions, the main shaft was pinned at both the upper and the lower ends of the shell, and in the analysis, the periphery of the tray and the inner wall of the shell were treated as contact pairs.

Figure 10a presents the von Mises stress distribution of the main shaft. It shows that the largest stress located at the connection between the shaft and the first module of the tray, which should be attributed to the higher pressure upon the tray from the top and the boundary condition at the upper end. It also reveals that the stress distribution in the middle of the shaft was homogeneous and the stress in the major bulk of it was still under the yield strength. However, as shown in Figure 10b, deflection did occur on the main shaft under the impact of the fluids, and the maximum radial deflection ranged from −5.8 to 7.8 mm. Obviously, it had partly exceeded the clearance (7.5 mm) and, consequently, induced wear on the shell. Furthermore, the result indicated that the maximum deflection was located at the same height with all the four modules of the tray, which accorded with the phenomenon observed in Figure 7 that the wear extent on the shell was of average severity along the vertical direction.

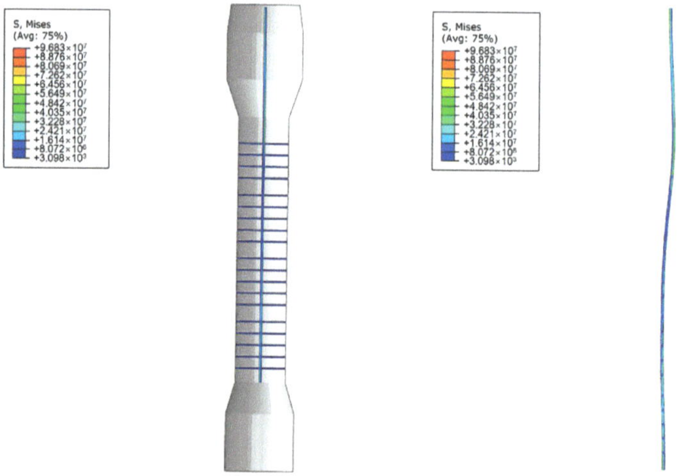


(**a**) von Mises stress distribution

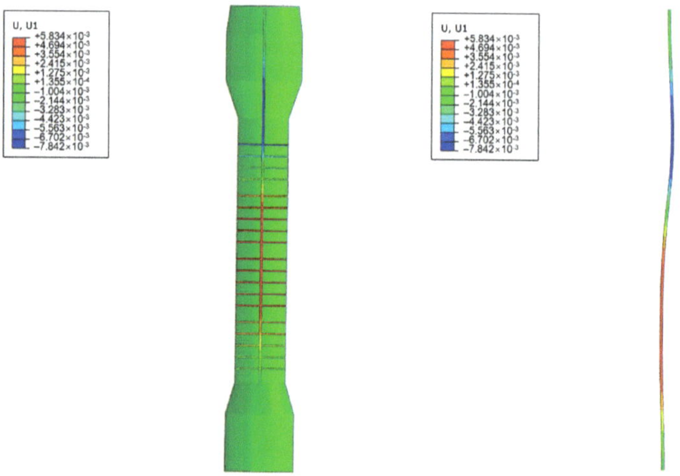


(**b**) deflection distribution

**Figure 10.** Finite element analysis results of the main shaft.

Since structural stiffness is the ratio between load and maximum deflection, as presented in Equation (2), it was determined that the main shaft with the current diameter did not have sufficient stiffness based on the analysis results. Then a new problem emerges of what is the minimum required value? Thus, further investigation was carried out to obtain the relation between the diameter of the main shaft and the largest deflection under the same loading condition.

$$K = \frac{P}{\delta} \quad (2)$$

where $K$ is the structural stiffness (N/m), $P$ is the load (N), $\delta$ is the maximum deflection (m).

As shown in Figure 11, without the limitation of the shell, the maximum deflection of the main shaft decreases monotonically with its diameter. In other words, the structural stiffness will increase with the diameter of the main shaft. Quantitatively, the maximum deflection can be reduced to 7.5 mm when the diameter reaches 105 mm. That is, if the current operational condition cannot be changed, 105 mm should be the minimum required diameter of the main shaft for prevention of deflection and subsequent wear upon the shell. To sum up, it could be proposed from the finite element analysis results that the inappropriateness of the current design was the root cause for abnormal wear on the shell.

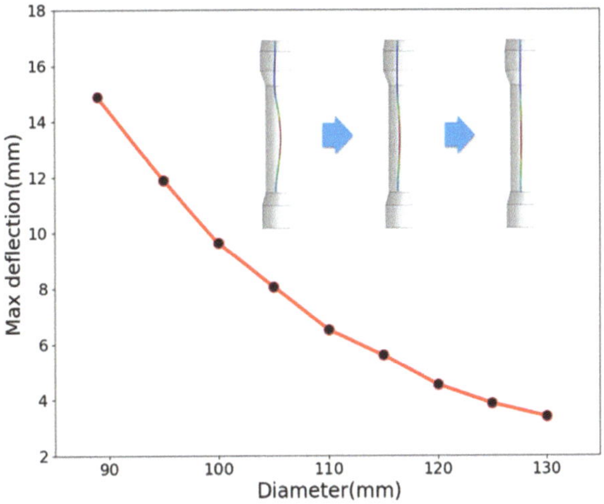


**Figure 11.** Maximum deflection versus the diameter of the main shaft.

### 3.3. Fabrication Safety Analysis

After analysis of the main shaft, now it comes to the tray. During the visual inspection, it was found that the shape of the sieve plates and the PTFE baffles was actually not perfectly round in general, seen in Figure 12a. That is, there existed some ovality. Furthermore, from the top view of the tray (Figure 12b,c), it was observed that the periphery of the sieve plates was partly damaged after wear; while for the PTFE baffles, the differences between their diameters even reached two to three centimeters, as presented in Figure 12d.

These phenomena clearly pointed one of the causes of wear to limited quality inspection of fabrication of the sieve plates and the PTFE baffles. Although the extraction function of the tray does not depend on the dimensional precision of the baffles and the sieve plates and a 7.5 mm-wide clearance is particularly reserved between them and the shell, the wear extent upon the shell will be aggravated due to the ovality of these two components once wear is initiated from deflection of the main shaft. In brief, although only ovality of the baffles and the sieve plates themselves would not result in wear on the shell of the extraction column, it was indeed an added factor for such severe wear due to limited quality inspection of fabrication.

In addition to ovality, another defect we discovered on the sieve plates and the PTFE baffles was warpage, as seen in Figure 13a,b. Especially for the baffles, part of the plate stacks and the PTFE gasket had already separated (Figure 13c), and the gasket had also suffered severe wear (Figure 13d).

It is not difficult to imagine that the warpage of such two components will narrow their clearance between the shell and, consequently, further aggravate the wear extent upon the shell. With regards to the causes of warpage, it was inferred that during the process of installation, the bolts for fixing the sieve plates and the PTFE baffles were assembled forcibly, leaving excessive tightening force on the two components. Thus, it was determined that limited quality inspection of installation was also an added cause for severe wear on the shell.

(**a**) total morphology from the bottom view

(**b**) damaged periphery of the sieve plates and PTFE baffles after wear

(**c**) damaged periphery of the sieve plates and PTFE baffles after wear

(**d**) diameters gaps of the PTFE baffles

**Figure 12.** External appearances of the tray.

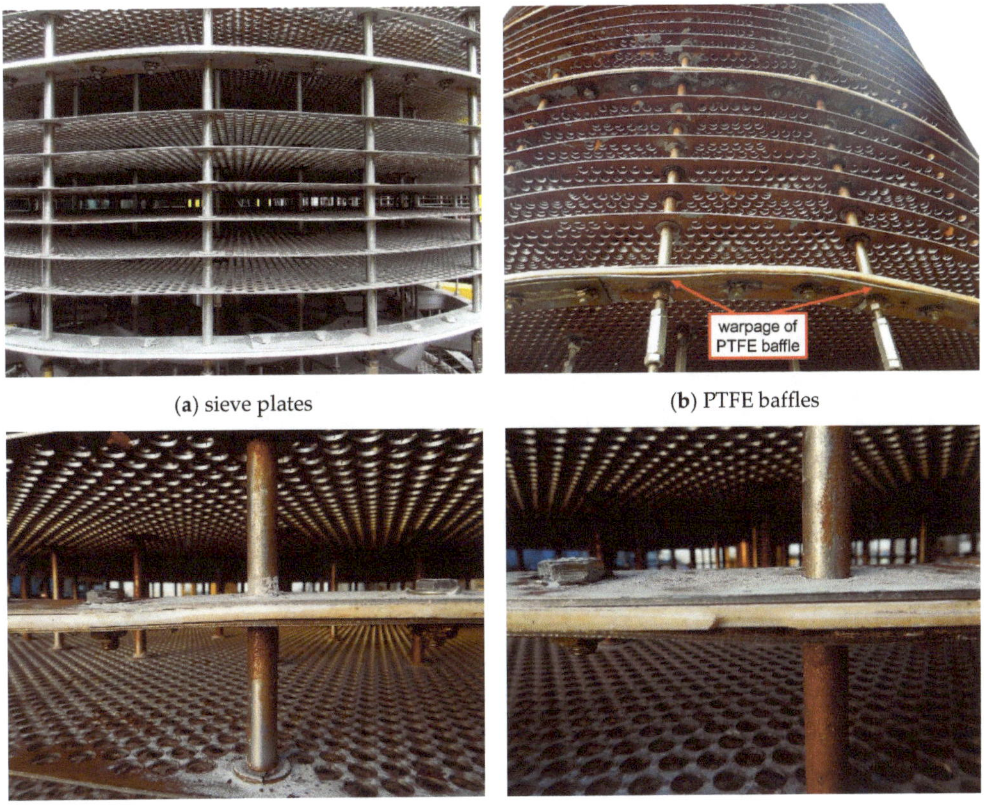


(**a**) sieve plates (**b**) PTFE baffles

(**c**) magnification of the PTFE baffles (**d**) magnification of the PTFE baffles

**Figure 13.** External appearances of warpage on the sieve plates and the PTFE baffles.

### *3.4. Effectiveness Verification*

Thus far, it is clear that insufficient structural stiffness of the main shaft due to inappropriate design was the root cause of deflection of the main shaft, which subsequently induced abnormal wear on the inner wall of the extraction column shell. In addition, ovality and warpage from limited quality inspection of fabrication and installation of the sieve plates and the PTFE baffles further aggravated the extent of wear. With regards to the types of wear, adhesive wear and fatigue wear should be mainly ascribed to, whose detailed mechanisms could be referred to as classic monographs [19] and would not be repeatedly discussed in this paper. However, coincidently, it can be summarized that all the three causes occurred in the phases prior to operation of the extraction column, i.e., design, manufacturing and installation. By the way, this wear phenomenon was also discovered in the other standby extraction column in the next routine overhaul, but the extent of it was not as severe because it had only served for one operation cycle.

Next is addressing the countermeasures. As for insufficient structural stiffness of the main shaft, on the basis of the finite element analysis results, to replace it with a new one with a diameter larger than 105 mm will certainly take effect. However, it will undoubtedly be timely and expensive, which is not acceptable for the company. Thus, the focus was then shifted to the measures that would not change the current service conditions of the extraction column and the current dimension of the main shaft. In other words, the measures should take into full consideration the fact that deflection of the main shaft would be inevitable. Then, we considered taking action from the other side. As indicated in Figure 11, the maximum deflection of the main shaft with a current diameter of 88.9 mm is about 15 mm, twice the clearance (7.5 mm) between the PTFE baffles and the shell of the column. Therefore, why not enlarge the clearance to tolerate the deflection? Thus, we suggested reducing the diameters of the sieve plates and the PTFE baffles by 5 to 10 mm. This time, it was accepted and immediately implemented by the company, as shown in Figure 14. From then on (over three years have passed), the problem of abnormal wear was no longer reported.

(**a**) polishing with grinding wheel (**b**) after polishing

**Figure 14.** Field personnel polishing the periphery of the sieve plates and the PTFE baffles.

In terms of the other two causes, including ovality and warpage of the sieve plates and the PTFE baffles due to limited quality inspection of fabrication and installation, the suggestions were relatively simple and directly accepted by the company, i.e., improving the quality of manufacturing and installation to eliminate ovality and warpage, respectively. Of course, they would only be beneficial to the extraction columns built in the future.

## 4. Conclusions and Countermeasures

Failure analysis was carried out for identification of the causes of abnormal wear on the inner wall of the extraction column shell, and the results pointed to the three phases prior to operation, which vividly demonstrated the 'Safety Design' concept for engineering equipment:

(1) In the design phase, the diameter of the main shaft was not large enough to provide necessary structural stiffness, so wear was induced due to deflection of the shaft.
(2) In the fabrication phase, the quality of the sieve plates and the PTFE baffles was not sufficiently inspected; therefore, ovality was left and aggravated the wear extent on the shell.
(3) In the installation phase, the warpage was formed on the sieve plates and the PTFE baffles because of limited quality inspection, which also aggravated the extent of wear.

Based on the above conclusions, countermeasures were put forward point by point for failure prevention of this extraction column:

(1) The periphery of the sieve plates and the PTFE baffles should be polished to reduce their diameters for tolerance of the main shaft deflection.
(2) The quality of manufacturing and installation of the sieve plates and the PTFE baffles must be improved for the elimination of ovality and warpage.
(3) Although no faulty operation from the field staff was found this time, training and management still need to be further strengthened, e.g., when abnormal signals are detected during the operation phase in the future, the equipment should be immediately stopped for root causes analysis.

**Author Contributions:** Resources and writing—original draft preparation, S.-H.W.; methodology and writing—review and editing, Y.G.; finite element analysis, Q.T. and J.-H.S.; investigation, X.-L.Y.; supervision and project administration, Z.-G.Y. All authors have read and agreed to the published version of the manuscript.

**Funding:** This research received no external funding.

**Institutional Review Board Statement:** Not applicable.

**Informed Consent Statement:** Not applicable.

**Data Availability Statement:** Not applicable.

**Conflicts of Interest:** The authors declare no conflict of interest.

## References

1. Ardente, F.; Beccali, G.; Cellura, M.; Lo Brano, V. Life cycle assessment of a solar thermal collector. *Renew. Energy* **2005**, *30*, 1031–1054. [CrossRef]
2. Schuman, C.A.; Brent, A.C. Asset life cycle management: Towards improving physical asset performance in the process industry. *Int. J. Oper. Prod. Manag.* **2005**, *25*, 566–579. [CrossRef]
3. Yang, D.Y.; Frangopol, D.M. Life-cycle management of deteriorating civil infrastructure considering resilience to lifetime hazards: A general approach based on renewal-reward processes. *Reliab. Eng. Syst. Saf.* **2019**, *183*, 197–212. [CrossRef]
4. International Atomic Energy Agency. *Ageing Management and Development of a Programme for Long Term Operation of Nuclear Power Plants*; IAEA Specific Safety Guide No. SSG-48; International Atomic Energy Agency: Vienna, Austria, 2018.
5. Kidam, K.; Hurme, M. Origin of equipment design and operation errors. *J. Loss Prev. Process Ind.* **2012**, *25*, 937–949. [CrossRef]
6. Yang, X.-L.; Gong, Y.; Yang, Z.-G. Failure analysis on abnormal leakage between tubes and tubesheet of spiral-wound heat exchanger for nuclear power plant. *Eng. Fail. Anal.* **2020**, *118*, 104900. [CrossRef]
7. Gong, Y.; Yang, Z.G. Fracture failure analysis of automotive accelerator pedal arms with polymer matrix composite material. *Compos. Part B Eng.* **2013**, *53*, 103–111. [CrossRef]
8. Gong, Y.; Ma, F.-Q.; Xue, Y.; Jiao, C.-S.; Yang, Z.-G. Failure analysis on leaked titanium tubes of seawater heat exchangers in recirculating cooling water system of coastal nuclear power plant. *Eng. Fail. Anal.* **2019**, *101*, 172–179. [CrossRef]
9. Gong, Y.; Du, M.-Y.; Ma, F.-Q.; He, G.-Q.; Xue, Y.; Jiao, C.-S.; Yang, X.-L.; Li, R.-B.; Liu, X.-Q.; Xu, X.-L.; et al. Failure analysis and prevention of corrosion occurring during storage on steam generator tube sheet for advanced PWR, Part I: Root causes analysis. *Eng. Fail. Anal.* **2020**, *116*, 104710. [CrossRef]
10. Catelani, M.; Ciani, L.; Venzi, M. Failure modes, mechanisms and effect analysis on temperature redundant sensor stage. *Reliab. Eng. Syst. Saf.* **2018**, *180*, 425–433. [CrossRef]
11. Deyab, S.M.; Taleb-berrouane, M.; Khan, F.; Yang, M. Failure analysis of the offshore process component considering causation dependence. *Process Saf. Environ. Prot.* **2018**, *113*, 220–232. [CrossRef]
12. Liu, A.; Chen, K.; Huang, X.; Chen, J.; Zhou, J.; Xu, W. Corrosion failure probability analysis of buried gas pipelines based on subset simulation. *J. Loss Prev. Process Ind.* **2019**, *57*, 25–33. [CrossRef]
13. Ciarapica, F.; Bevilacqua, M.; Antomarioni, S. An approach based on association rules and social network analysis for managing environmental risk: A case study from a process industry. *Process Saf. Environ. Prot.* **2019**, *128*, 50–64. [CrossRef]
14. Gong, Y.; Yang, C.; Yao, C.; Yang, Z.G. Acidic/caustic alternating corrosion on carbon steel pipes in heat exchanger of ethylene plant. *Mater. Corros.* **2011**, *62*, 967–978. [CrossRef]
15. Fei, J.-L.; Yang, X.-L.; Wang, S.-H.; Gong, Y.; Wang, J.; Tao, X.; Yang, Z.-G. Failure analysis and countermeasures for localized abnormal wear of MAA extraction column. *Eng. Fail. Anal.* **2019**, *100*, 166–179. [CrossRef]
16. Zhang, J.; Zhang, L.; Liang, Z. Buckling failure of a buried pipeline subjected to ground explosions. *Process Saf. Environ. Prot.* **2018**, *114*, 36–47. [CrossRef]
17. Gong, L.-H.; Zhu, W. Corrosion failure analysis of a perforated F32 reactor. *Mater. Test.* **2019**, *61*, 533–536. [CrossRef]
18. Sulu, I.Y.; Temiz, S. Failure analysis of an adhesively joined composite pipe system under internal pressure. *Mater. Test.* **2018**, *60*, 997–1003. [CrossRef]
19. Blau, P.J. *ASM Handbook Volume 18: Friction, Lubrication, and Wear Techonology*; ASM International: Geauga, OH, USA, 1992.

*materials*

*Article*

# Numerical Failure Analysis and Fatigue Life Prediction of Shield Machine Cutterhead

**Jie Li *,†, Zengqiang Zhang †, Chuang Liu, Kang Su and Jingbo Guo**

School of Mechanical Engineering, Shijiazhuang Tiedao University, Shijiazhuang 050043, China; zhangzq1127@stdu.edu.cn (Z.Z.); l_c19@126.com (C.L.); sukang@stdu.edu.cn (K.S.); guojingbo66@163.com (J.G.)
* Correspondence: lijie@stdu.edu.cn; Tel./Fax: +86-311-8793-5228
† These authors contributed equally to this work and should be considered co-first authors.

**Abstract:** This paper presents numerical failure analysis on cracking of shield machine cutterhead structure during a metro-tunnel construction. The stress intensity factors (SIFs) of surface cracks with different shapes and location angles were analyzed by a finite element simulation method based on linear elastic fracture mechanics (LEFM) theory. The ratios of variation in stress intensity factors of cracks with different shapes were analyzed. The maximum allowable crack depth of the cutterhead panel is 50.23 mm by dynamic stress calculation, and the damage tolerance criterion of the cutterhead panel was proposed. The influence of the Paris model parameter values was analyzed based on mathematical methods. It is proven that the location of the cutterhead cracking angle is mainly determined by the mixed-mode SIF. In practice, the crack section basically expanded into the semi-elliptical shape. The cutterhead structure may directly enter the stage of crack propagation due to welding defects during tunneling. The research results provide a theoretical basis and important reference for crack detection in the key parts of the cutterhead, as well as maintenance cycle determination and life prediction of the cutterhead mileage, both of which have important engineering value.

**Keywords:** cutterhead; failure analysis; life prediction; crack propagation; stress intensity factor

**Citation:** Li, J.; Zhang, Z.; Liu, C.; Su, K.; Guo, J. Numerical Failure Analysis and Fatigue Life Prediction of Shield Machine Cutterhead. *Materials* **2021**, *14*, 4822. https://doi.org/10.3390/ma14174822

Academic Editor: Yi Gong

Received: 27 June 2021
Accepted: 23 August 2021
Published: 25 August 2021



## 1. Introduction

The cutterhead of a shield machine is generally a large welded structure, which is welded using steel plate to form a whole structure, and the corresponding position is reserved to install the disc cutters and scrapers. The structure of a shield machine is shown in Figure 1.

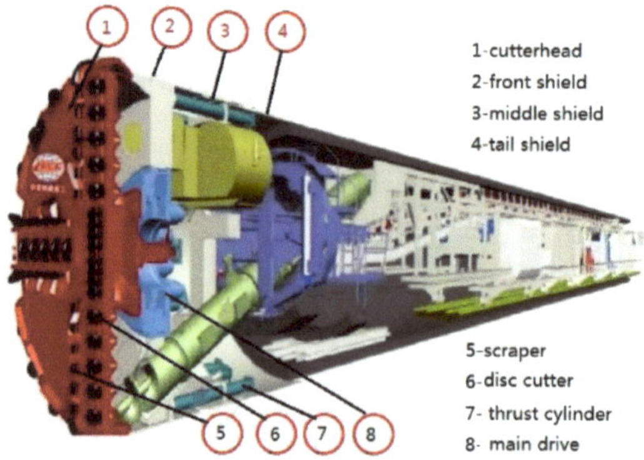


**Figure 1.** Structure of shield machine.

The working principle of a cutterhead is as follows: the cutterhead breaks and cuts the soil using the cutters installed on the panel with the comprehensive action of thrust and torque; the tunnel section is formed once. Generally, the cutterhead of a shield machine is welded by carbon-dioxide gas shielded welding, and an ultrasonic testing method is used for flaw detection, with the highest accuracy being Φ2 mm. Due to the extremely complex geological conditions in the process of cutterhead excavation, cracks will be initiated and propagated in small defects or weak links by the action of alternating loads, which can result in structural failure. The wear and fatigue cracks are the main forms of cutterhead failure, as shown in Figure 2.

**Figure 2.** Failed cutterhead of a shield machine.

The cutterhead is the core component of a shield machine. Its service performance directly affects the excavation efficiency of a shield machine and the safety of the construction. Considering the complex structure of a shield machine cutterhead, the problem of reliability becomes particularly important. The static design method is predominantly used in the cutterhead structure design, and a large safety factor is involved to ensure the safety of the cutterhead structure. In the construction of hard strata, such as bedrock or solitary stone, the problems of excessive wear and fatigue cracking of the cutterhead appear easily, endangering the safety of tunneling construction.

In the fatigue life prediction of large structures such as cutterhead, Ling [1] proposed a large-scale structure life prediction method based on system dynamics, linear elastic fracture mechanics and fatigue damage accumulation theory. The predicted cutterhead driving mileage is able to meet the design objectives. Sun Wei et al. [2] studied the cutterhead panel material Q345 multi-crack propagation damage law and gave the change rule, and propagation path of stress intensity factors during the propagation process of collinear crack, parallel crack and non-parallel crack. The results lay a foundation for predicting the multi-crack propagation life of a cutterhead. Ling & Sun et al. [1,3] established the probability density function of the load distribution of a cutterhead using statistical method, and then calculated it using a rain flow counting method, and compiled the eight-level load spectrum. In the fatigue crack-growth calculation model, Liu et al. [4] put forward the small-time scale prediction model based on continuous crack-growth, and obtained better prediction results for aluminum alloy and other materials. Huo et al. [5] predicted the influence of thick plate on crack-growth behavior, based on the improved small-time scale model, and obtained the expression of constraint factor function. Due to the complex structure of a cutterhead and changeable geological environments, the cracks generated by a cutterhead are mostly composite cracks [1,3,6]. The stress intensity factors of mode I, II and III of mixed-mode cracks are different, and are relatively complex crack forms.

Dicecco Sante et al. [7] studied the effect of surface corrosion on the high-cycle fatigue and low-cycle fatigue of Q345 mining wheel. When the test cycle reached $5 \times 10^5$, the fatigue strength of the corroded surface decreased by 24.6% compared with the low-cycle fatigue. At the same time, the fracture behavior of the low-cycle fatigue corrosion sample and the polished sample were obviously different. Dong et al. [8] Studied the low-cycle fatigue mechanism of Q345 steel for pressure vessel, established the linear relationship between crack-tip opening displacement and crack-growth rate, and analyzed the influence of the plastic zone. Zong et al. [9] studied the fatigue crack-growth rate of bridge steel Q345qD. Based on probability statistics, the mean value and variance estimation model of parameter C and m of the Paris formula for crack-growth and the design parameters were given. The results can provide reference for the fatigue resistance design of steel structures. The Paris model was put forward by Paris and Erdogan [10] in 1963, and was further developed and improved later by many scholars. The landmark theories and formulas are Austen growth model [11], Forman model [12] and Neuman [13] crack-tip opening displacement model to calculate the crack-growth rate. Some scholars have developed the Paris model and applied it to aluminum alloy [14] and thick plate [5] and welding joint [15]. These studies enrich the application range of the Paris formula and provide a more accurate model to predict the structural fatigue life.

Some scholars [8,9] obtain material crack-growth parameters C and m by means of experiments and data statistics method. These data take material homogeneity, experimental repeatability and statistical reliability into account, and can be used as basic data for structural design and evaluation. Generally, the parameter m has a greater influence on the fatigue crack-growth life [3]. According to observations made in experiments conducted by researchers, the results of the average value analysis of general statistics are more general and representative.

Some scholars have studied the shear propagation of cracks in materials. Feng Yu et al. [16] presents an experimental study on diagonal crack width estimation of Shear-Strengthened Pre-damaged Beams with CFRP strips (SSPBCs). Several parameters including pre-damaged degree, shear-span ratio and CFRP strips spacing are considered. The crack formation of shear-strengthened undamaged or low pre-damaged beams with CFRP strips is caused by reaching the ultimate tensile strain of concrete, while that of shear-strengthened high pre-damaged beams with CFRP strips is due to the relative slip between stirrups and the concrete. The development rate of diagonal crack increases as the shear-span ratio, CFRP strips spacing, or pre-damaged degree increases. Yuya Tanaka et al. [17] investigated the shear-mode crack-growth for the fatigue strength of Ni-base superalloy. Three different types of fatigue tests were performed: (i) push-pull; (ii) pure-torsion; (iii) torsion with superposed static tension. All tests revealed non-propagation of small, shear-mode cracks.

In summary, some progress has been made in the research on large structure fatigue life prediction, and the crack fatigue life under service conditions has been predicted. The general and universal calculation flow of different geological and cutterhead forms are still lacking. Under different structures of cutterhead and different geological conditions, it is necessary to analyze the failure of the cutterhead and conduct the life prediction of the cutterhead.

A shield machine was used in the construction of a subway tunnel in Xuzhou of China. The cross-section of the tunnel is a composite stratum, and there are bad strata such as bedrock intrusion; the maximum value of uniaxial compressive strength is 122 MPa. When the tunnel section is perforated, cracking is found in the front panel, as shown in Figure 3. The crack length is about 1.25 m, and the crack depth is about 16 mm. One side is located at the welding site of the cutterhead beam and scraper (zone 2#), and the other side is located at the center cutter saddle (zone 1#). The crack is basically a straight line. The angle with the transversal direction is approximately about 53~60°.

To solve this problem, numerical failure analysis was carried out to identify the root causes, and the crack-growth law was applied to predict the crack-growth life of the cutterhead. This paper demonstrates an engineering case of structural failure due to fatigue crack on a certain type of cutterhead, which ensures the safety of the construction process and provides reference for similar structures.

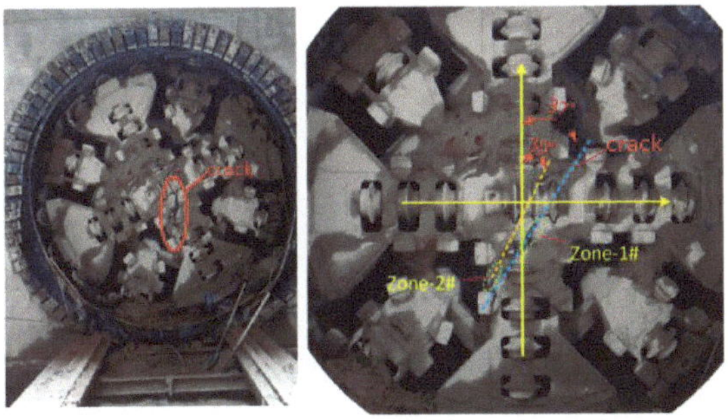


**Figure 3.** Diagram of cutterhead cracking.

## 2. Modeling and Static Analysis of Shield Cutterhead

### 2.1. 3D Modeling of Cutterhead

The 3D model of shield cutterhead is established in SolidWorks according to the ratio of 1:1. Without affecting the overall structural accuracy, the model of cutterhead is simplified appropriately. The characteristics of bolt hole, chamfer and rounded corner are omitted, and the weld seam is rigidized. The cutterhead model is imported into ANSYS for analysis. The overall structure of the cutterhead is welded into a whole, as shown in Figure 4. The outer side of the cutterhead is welded with scrapers. The opening rate of the cutterhead is about 30%. The rear side of the cutterhead is connected with four supporting ribs to form a whole. The bracket is connected with the back flange. The detailed parameters of the cutterhead are shown in Table 1.

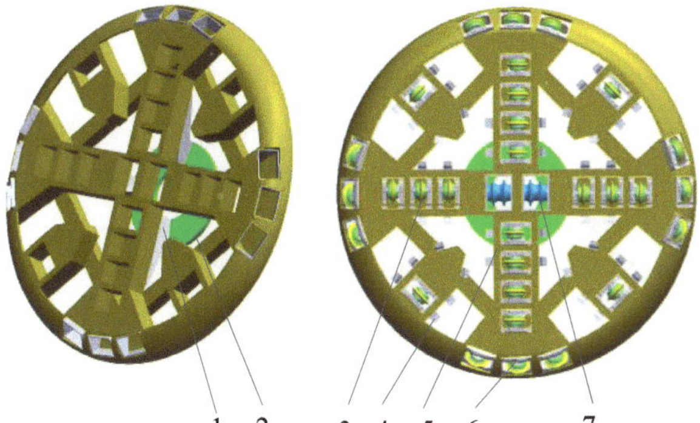


**Figure 4.** Model of shield cutterhead. 1-support; 2-flange; 3-normal disc cutter; 4-Scraper; 5-cutter beam; 6-gage disc cutter; 7-central disc cutter.

**Table 1.** Cutterhead profile.

| Main Technical Parameters of Cutterhead | |
|---|---|
| Excavation diameter/mm | Φ 6280 |
| Cutterhead material | Q 345 D |
| Total weight/t | About 75 |
| Number of 17-inch single-edged cutters diameter $d_0$/mm | 30/Φ 432 |
| Number of 19-inch double-edged cutters diameter $d_1$/mm | 2/Φ 483 |
| Scraper | 40 |

### 2.2. Static Analysis of Cutterhead

Through static analysis of the cutterhead structure, the distribution of stress-strain of the cutterhead structure is obtained. The position of the maximum stress is sought, which is set as a dangerous point. Static analysis lays a foundation for further fracture mechanics analysis.

The whole cutterhead is imported into ANSYS Workbench and meshed by Tetrahedrons element. The mesh size is adaptive and the accuracy is medium. The number of elements is 27,742, and the number of nodes is 54,060. The average element size is 60 mm. The cutterhead boundary conditions are set as follows: the flange is added with fixed constraints, and the surface force is 0.80 MPa and the torque is 2000 N·m. The calculated results are shown in Figure 5. The figure shows that the maximum deformation of the cutterhead is 3.840 mm and the maximum stress of the cutterhead is 190.67 MPa. The cutterhead safety factor $s_1$ is 1.81, which indicates the structural static safety. In Figure 5, it can be seen that the local maximum stress is 190.67 MPa, which is located on the inner side of the cutterhead beam support plate. The cutterhead structure can be simplified to beam structure, and it is easy to produce stress concentration, which is also the focus of cutterhead maintenance.

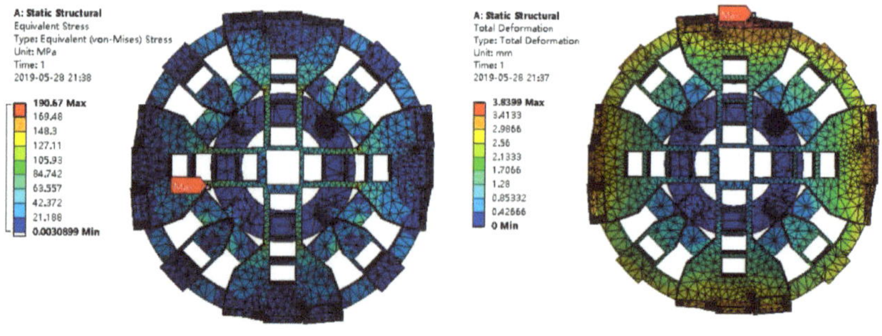


**Figure 5.** Static analysis results of cutterhead.

## 3. Crack Modeling and Analysis

According to the linear elastic fracture mechanics theory, a semi-elliptical surface crack is inserted at the initial cracking location. The numerical method [1] is used to solve the stress intensity factor, and the crack propagation variation law is analyzed according to the variation law of the stress intensity factor.

### 3.1. Analysis of Cracking Direction

The crack began on zone 2# of the cutterhead of the shield machine during tunneling. After initiation, the crack entered the stage of propagation and expanded continuously along the directions of length and depth.

A schematic diagram of the crack position angle is illustrated in Figure 6. $xoy$ is the absolute coordinate system, $x_L$ is the direction of the crack length, $\theta$ is the angle between the crack length and the $x$-axis, and the crack length is $2c$. The finite element model of the cutterhead with the same crack size ($2c = 60$ mm, $a = 15$ mm) was taken as the research object under the maximum loading condition. Considering the randomness of crack position angles, the variation rules of crack stress intensity factor at different crack position angles were analyzed. The variation law of the stress intensity factors of the cracks was obtained. The stress intensity factors of the three cracks in the range of 0~90° were calculated by taking 15° as interval.

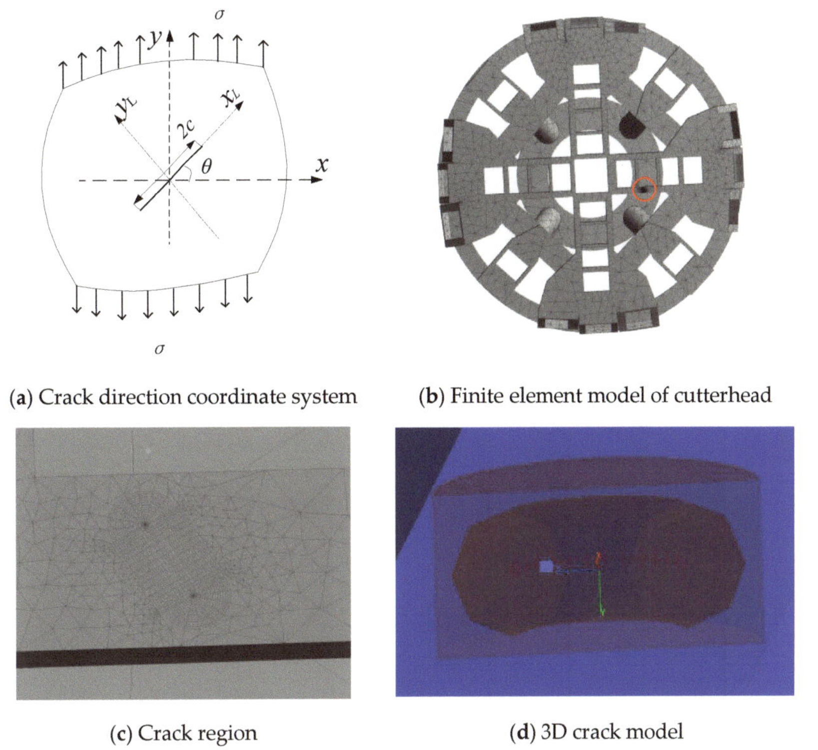


(**a**) Crack direction coordinate system (**b**) Finite element model of cutterhead

(**c**) Crack region (**d**) 3D crack model

**Figure 6.** Definition of crack position angle: (**a**) Crack direction coordinate system (**b**) Finite element model of the cutterhead (**c**) Crack region (**d**) 3D crack model.

As shown in Figure 7, crack stress intensity factors of different crack centrifugal angles under different crack position angles are presented.

Figure 7 demonstrates the following conclusions:

(1) The stress intensity factors of mode I are basically distributed symmetrically. With the increase of $\theta$ angle, the value of stress intensity factors decreases, and tends to be flat near 90°. When $\theta = 0°$, the value of stress intensity factor reaches the maximum 414.87 MPa·mm$^{1/2}$. When $\theta = 90°$, the value of stress intensity factor is negative. The stress intensity factors of mode I only exists when it is open. When $K_I < 0$, it has no significance.

(2) The stress intensity factors of mode II are basically central-symmetric with $\theta = 90°$, the value decreases from left to right, and the overall value increases first and then decreases with $\theta$ angle. When $\theta = 45°$, the maximum value is 197.88 MPa·mm$^{1/2}$.

(3) The stress intensity factors of mode III are basically symmetrical. The absolute values of the stress intensity factors increase first, and then decrease, with the increase of $\theta$ angle. Except for $\theta = 0°$ all the stress intensity factors are positive, indicating that the increase of $\theta$ angle changes the tearing direction, and the maximum value is 179.49 MPa·mm$^{1/2}$ at $\theta = 45°$.

The main driving force of the crack propagation direction comes from the stress intensity factor at the free ends of the crack surface [18]. As shown in Figure 8, the stress intensity factors of mode I and II are always more than two times of mode III in the range from 0~75°. Angle ranges from 75~90° are about twice as much. From the comprehensive analysis, it can be seen that the main form of crack cracking is the mixed mode.

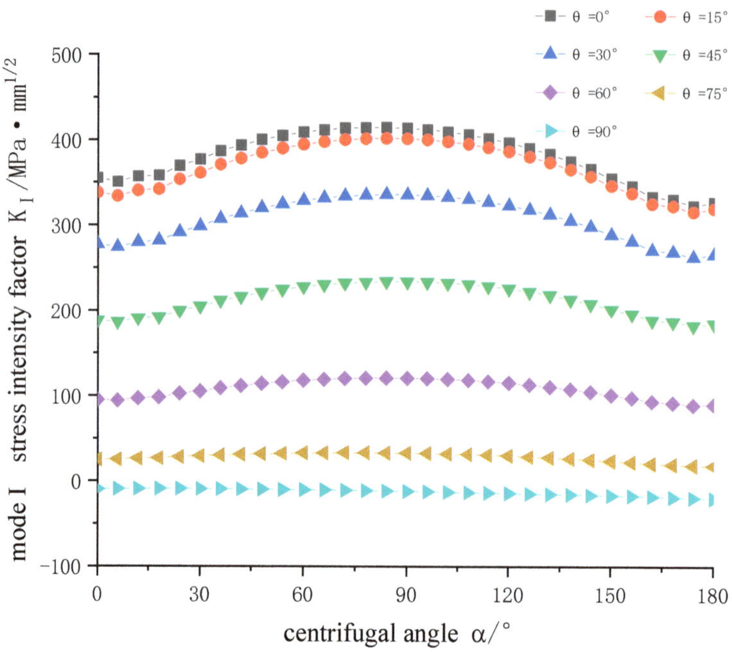


(**a**) Mode I stress intensity factor distribution

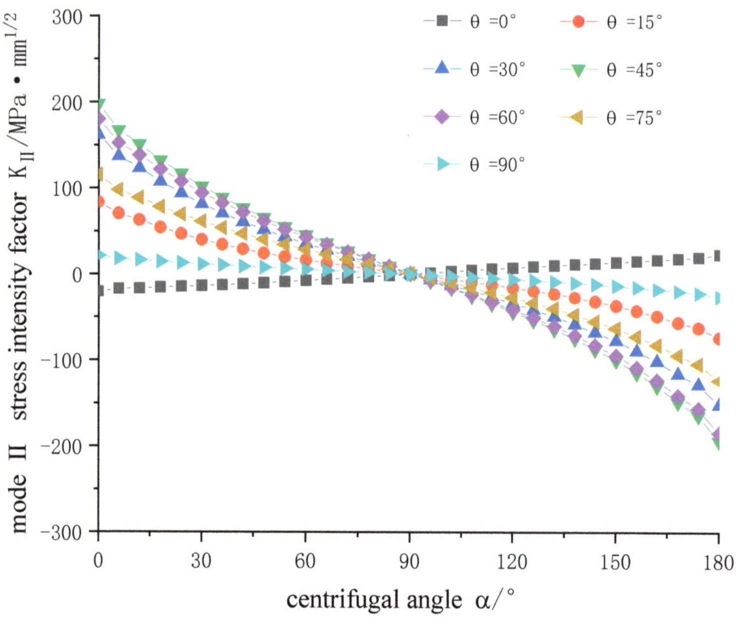


(**b**) Mode II stress intensity factor distribution

**Figure 7.** *Cont.*

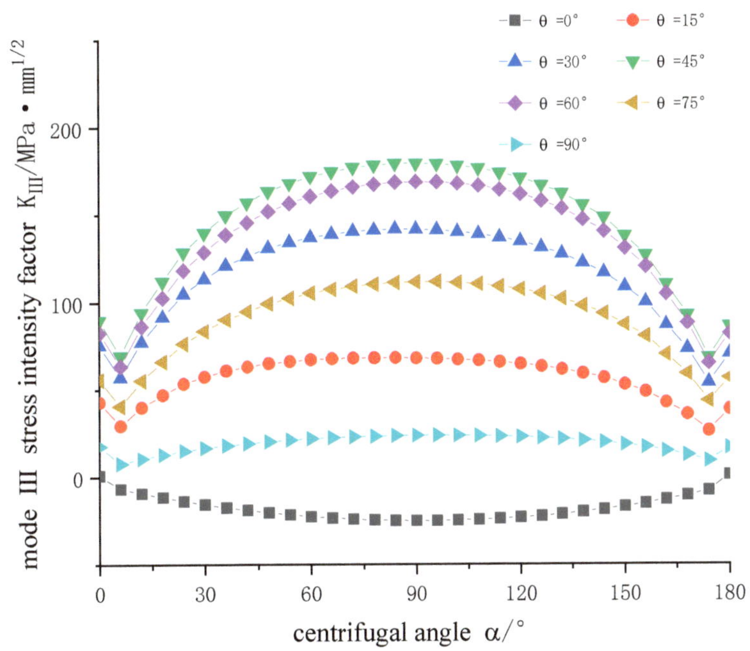


(**c**) Mode III stress intensity factor distribution

**Figure 7.** Three modes stress intensity factor distributions: (**a**) Mode I stress intensity factor distribution (**b**) Mode II stress intensity factor distribution (**c**) Mode III stress intensity factor distribution.

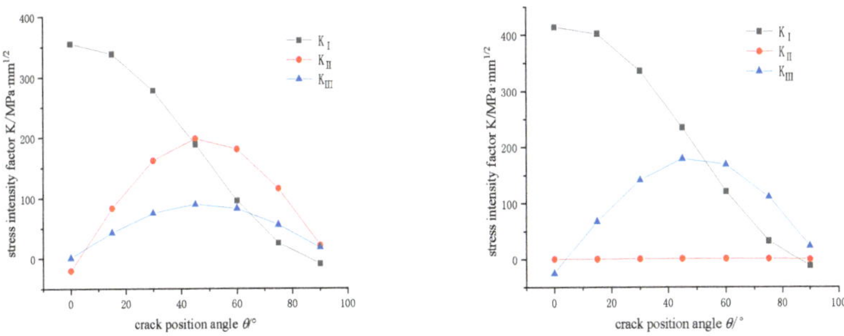


(**a**) SIFs distribution at free ends of crack (**b**) SIFs at the deepest crack

**Figure 8.** Stress intensity factor distribution of cracks: (**a**) SIFs distribution at free ends of crack (**b**) SIFs at the deepest crack.

The formula of maximum circumferential stress [19] in the polar coordinate system is established as

$$\begin{cases} \sigma_r = \frac{1}{2\sqrt{2\pi r}}\left[K_I(3-\cos\theta)\cos\frac{\theta}{2} + K_{II}(3\cos\theta-1)\sin\frac{\theta}{2}\right] \\ \sigma_\theta = \frac{1}{2\sqrt{2\pi r}}\cos\frac{\theta}{2}\left[K_I\cos^2\frac{\theta}{2} - \frac{3}{2}K_{II}\sin\theta\right] \\ \tau_{r\theta} = \frac{1}{2\sqrt{2\pi r}}\cos\frac{\theta}{2}\left[K_I\sin\theta + K_{II}(3\cos\theta-1)\right] \end{cases} \tag{1}$$

where, $\cos(\theta/2) \neq 0$ that is $\theta \neq \pm\pi$, and two free surfaces are not considered. The radius $r$ does not tend to zero, otherwise infinity will occur, which means the crack tip will not be considered. The condition for circumferential stress to extremum is

$$\frac{\partial \sigma_\theta}{\partial \theta} = 0 \tag{2}$$

The simplified formula is

$$K_I \sin\theta_0 + K_{II}(3\cos\theta_0 - 1) = 0 \tag{3}$$

When $K_I$ and $K_{II}$ values are introduced, the solution of $\theta_0$ can be obtained. From the actual crack direction, the crack position angle is about 53~60°. When $\theta = 53°$ is brought, $K_I$ = 138.66 MPa·mm$^{1/2}$ and $K_{II}$ = 192.61 MPa·mm$^{1/2}$, we obtained $\theta_0$ = 57.58°. The relative error of the solution is 8.6%. When $\theta = 60°$is brought, $K_I$ = 95.455 MPa·mm$^{1/2}$ and $K_{II}$ = 180.55 MPa·mm$^{1/2}$, and $\theta_0$ = 60.84°is obtained, the relative error is 1.4%. The main factors causing errors are the influence of mode III stress intensity factor $K_{III}$. Moreover, the closer the numerical value of the stress intensity factor is to the real crack-position angle, the smaller is the error.

### 3.2. Crack Propagation Law with Different Shape Ratio

The actual crack shape changes during the crack-growth process [20], which is due to the inconsistency of crack-growth rate in the depth and length directions. It is necessary to reveal the crack propagation behavior and analyze the distribution of crack stress intensity factors of cracks with different shapes.

First, a cutterhead model with a semi-elliptical crack is established at the zone 2#. The crack length axis c = 30 mm is taken as the same, the crack depth a is constantly changing. The shape ratio *a/c* are 0.1, 0.3, 0.5, 0.8 and 1.0, then the simulation model of the position angle 60° crack is sequentially established. The distribution of stress intensity factors at different crack shape ratios is analyzed and the results are shown in Figure 9.

From the stress intensity factors curves in Figure 9, insights are gained as follows:

(1) The stress intensity factors of mode I crack are basically symmetrical, and range from 50 to 150 MPa·mm1/2. With the increase of the crack shape ratio, the values of stress intensity factors increase gradually. When the crack is very shallow, the main propagation is depth growth, the length growth is secondary; when the crack approaches circular (a/c = 1.0), the stress intensity factor of crack is basically linear.

(2) The stress intensity factors of mode II crack are basically 90° center symmetry, indicating that the direction of crack slip has changed. As the shape ratio increases, the values of the stress intensity factor increase gradually. It shows that the closer to the circle, the faster the expansion speed.

(3) The stress intensity factors of mode III crack are symmetrical, increase firstly and then decrease. Except for the narrow crack of a/c = 0.1, the other crack stress intensity factors are close to each other, and the maximum value is near 150 MPa·mm1/2. When a/c = 0.8 and 1.0, the free ends of the crack appear singular, and the endpoint singular values are discarded. The crack-growth path is generally controlled and affected by many factors, which is one of the key research directions in the next stage. This paper offers some preliminary discussions. For example, at the zone #1, the cracks are mainly mode I cracks, so the overall crack-growth trend is linear, but the local path twists and turns under the control of mode II and mode III stress intensity factors, due to the structure and load. However, according to the law of crack propagation path, the corresponding crack arrest structure design can be carried out to prolong the structural life.

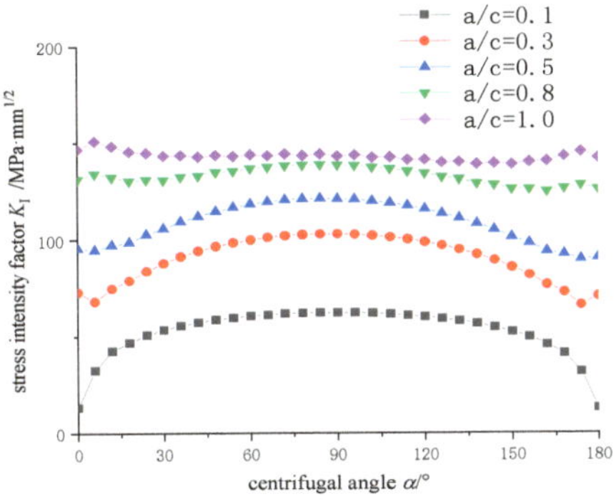


(**a**) Stress intensity factor of mode I crack

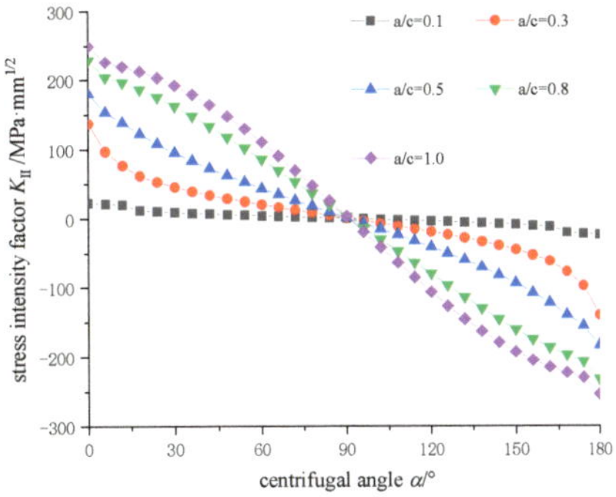


(**b**) Stress intensity factor of mode II crack

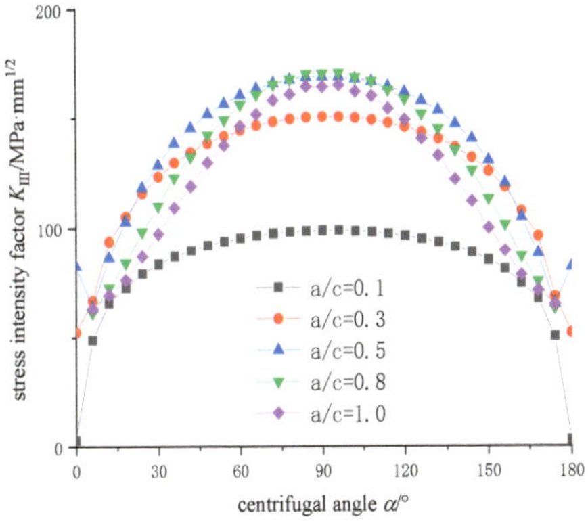


(**c**) Stress intensity factor of mode III crack

**Figure 9.** Stress intensity factor distribution of cracks with different shape ratio: (**a**) Stress intensity factor of mode I crack (**b**) Stress intensity factor of mode II crack (**c**) Stress intensity factor of mode III crack.

## 4. Crack Propagation Life of Cutterhead

### *4.1. Initial Crack Size Determination*

The material of a cutterhead is mainly Q345D steel, and the cutterhead is assembled as a whole structure through welding. The physical parameters of cutterhead material Q345D are shown in Table 2, where D indicates that the V-notch impact test energy of the material is greater than 27 J at −20 °C. Problems such as long weld length, plate thickness and difficult penetration may lead to welding defects. The sensitivity of the ultrasonic test (UT) instrument in engineering application is mostly $\phi$2 mm [3], which can detect the initial crack size of 2 mm or more. After a long period of heavy load, vibration and other comprehensive effects, the welding defects gradually expand until the strength of the cutterhead is insufficient and the fracture/failure occurs.

Initial surface crack size refers to the crack size that begins to calculate the life of the crack propagation stage, and can be evaluated by non-destructive testing. In applying engineering considerations, the crack size should be determined comprehensively on the basis of considering the allowable defect degree of structure, the accuracy of existing instruments and the technical level of operators. The initial crack depth used in engineering $a_0$ is 0.5 mm [4]. According to the conventional ultrasonic testing method, the crack length $2c_0$ is 2 mm.

**Table 2.** Physical parameters of Q345D material.

| Serial Number | Performance Index | Numerical Value |
|---|---|---|
| 1 | Density | 7850 $kg/m^3$ |
| 2 | Elastic modulus | 210 GPa |
| 3 | Poisson's ratio | 0.3 |
| 4 | Yield strength $f_y$ | 345 MPa |
| 5 | Ultimate tensile strength $f_u$ | 500 MPa |
| 6 | Breaking threshold $\Delta K_{th}$ | 201.12 $MPa{\cdot}mm^{1/2}$ |
| 7 | Fracture toughness $K_{IC}$ | 6270.8 $MPa{\cdot}mm^{1/2}$ |
| 8 | Thermal conductivity | 48 W/m·K |
| 9 | Coefficient of linear expansion | $1.2 \times 10^{-5}$ |
| 10 | Mass heat capacity | 480 $J{\cdot}m^{-1}{\cdot}K^{-1}$ |

### 4.2. Criterion of Crack Damage Depth Tolerance

The critical crack size is the allowable maximum crack size of a cutterhead structure, which is generally expressed by $a_c$. The critical crack size is determined by $K$ criterion, and thereby can be obtained as follows:

$$a_c = \frac{1}{\pi}\left(\frac{K_{IC}}{\alpha n \sigma_{max}}\right)^2 \tag{4}$$

In Formula (4), the crack shape coefficient $\alpha$ is 1.1. In reference [20], the safety factor $s_1$ is 2 and the Q345 fracture toughness value $K_{IC}$ is 203.08 $MPa{\cdot}m^{1/2}$. As seen in Figure 10, the maximum stress $\sigma_{max}$ is 232.38 MPa. When introducing the above values into Equation (4), the damage tolerance value of the cutterhead crack depth direction can be obtained as follows:

$$a_c = \frac{1}{\pi}\left(\frac{203.08}{1.1 \times 2 \times 232.38}\right)^2 \times 1000 = 50.23\text{mm}$$

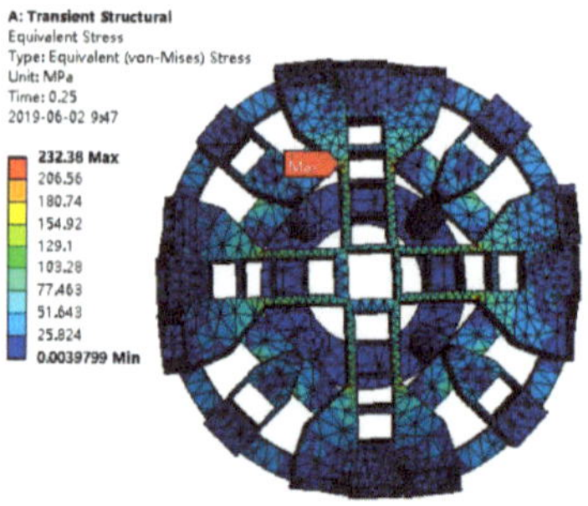


**Figure 10.** Transient dynamics nephogram.

The criterion of depth damage tolerance of the cutterhead panel is proposed, which provides a basis for further calculation of crack-growth life.

### 4.3. Analysis of Fatigue Crack-Growth Rate Parameters

In 1960′s, the Paris model [10] was established to calculate the crack-growth life. The results of previous research, references [3,10], show that the fatigue parameters $C$ and $m$ are dispersive. The fatigue parameters $C$ range from $1.0619 \times 10^{-13}$ to $3 \times 10^{-13}$ and $m$ range from 3.07 to 4.76.

Crack depth and length direction propagation rates are:

$$\begin{gathered} da/dN = C_a\left(\Delta K_{eqa}\right)^m \\ dc/dN = C_c\left(\Delta K_{eqc}\right)^m \\ C_c = 0.9^m C_a \end{gathered} \quad (5)$$

The crack propagates in both depth and length directions. According to Formula (5), the partial derivatives of $C_a$ and m are calculated, respectively, and the effect of the partial derivatives on the crack-growth rate is investigated. As shown in Figure 11, taking $m$ = 4.0 and $\Delta K_{eqa}$ = 50 MPa·mm$^{1/2}$, and the rate increases linearly with $C_a$ and exponentially with $m$. When $m$ less than 4.0, the rate changes relatively gently. The crack-growth rate increases rapidly with $m$ increasing when m is bigger than 4.0. When the stress intensity factors of the crack ranges are in the same interval, the depth growth rate is larger than the length expansion rate. It is recommended that C should be less than $2 \times 10^{-13}$ and $m$ should be less than 4.0, which tends to be safe.

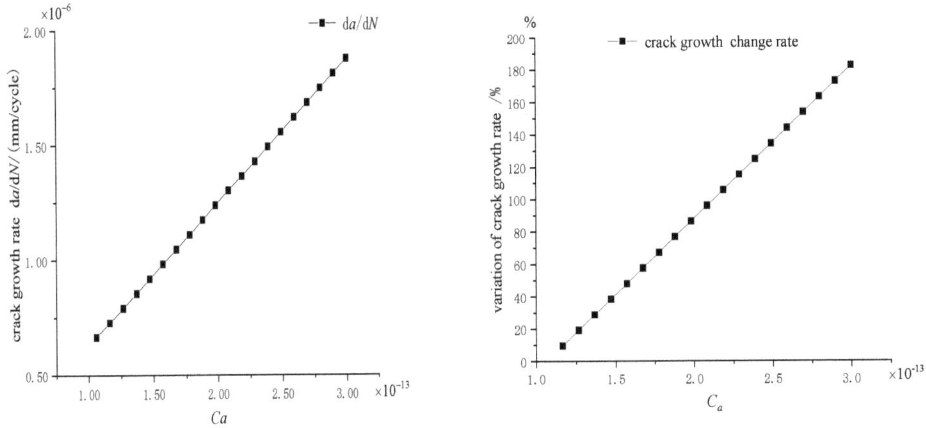


**(a)** Variation in Ca and crack-growth rate

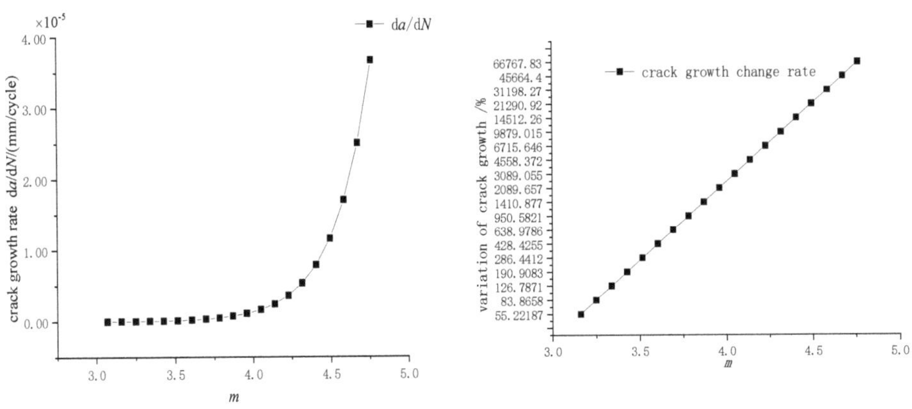


**(b)** Variation of m and crack-growth rate

**Figure 11.** The effect of parameters changes on the growth rate: (**a**) Variation in Ca and crack-growth rate (**b**) Variation in m and crack-growth rate.

## 4.4. Prediction Model of Crack Propagation Life

The cutterhead welding defects have experienced the process of crack initiation and expansion during the excavation process. Crack initiation life accounts for the vast majority of fatigue life, even more than 80%. Based on the conditions of the engineering site application, the cutterhead crack is generated during the tunneling process. The deepest crack stress intensity factor of the difference depth $a$ between $\Delta K_{eqa}$ drive crack is extended into depths. The cracks always keep shape with semi-ellipse [5]. The shape ratio $a_0/c_0 = 0.5$ was determined according to the initial crack. The stress intensity factor values of the crack at four different depths are calculated, and using the method of quadratic equation fitting, the curve is shown in Figure 12. Meanwhile the quadratic fitting formula (6) is obtained as follows, and the fitting coefficient $R^2$ is 0.969.

$$\Delta K_{eqa} = 135.562 - 4.483a + 0.509a^2 \tag{6}$$

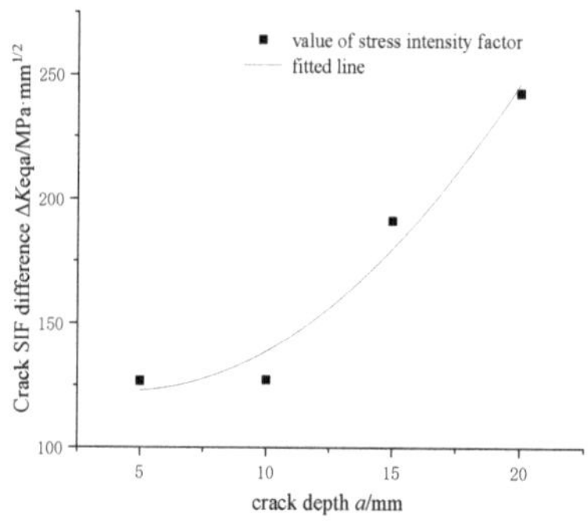


**Figure 12.** Fitting curve of crack stress intensity factor with crack depth.

According to reference [4,10], the calculation flow of cutterhead crack-growth life is presented as Formula (7) based on the Paris model.

$$N = \int_{N_0}^{N_c} \mathrm{d}N = \int_{a_0}^{a_c} \left(\frac{1}{\mathrm{d}a/\mathrm{d}N}\right)da = \int_{a_0}^{a_c} \frac{1}{C(\Delta K_{eqa})^m}\mathrm{d}a \tag{7}$$

Using the values for C as $2 \times 10^{-13}$ and m as 3.5, we then introduce them into Formula (7) to obtain the load cycle number $N$.

Each rotation of the disc cutter is applied as a cycle load excitation to the crack of the cutterhead. From the rotation center of the cutterhead, the crack position radius at the cutterhead $r_1$ is 800 mm. Assuming that the disc cutter rolls purely, the disc cutter rotation speed $n_c$ can be expressed as follows:

$$n_c = \frac{n \times r_1}{d_0/2} = \frac{n \times 800}{216} = 3.70n \tag{8}$$

The average rotation speed $n$ and penetration $p$ of the cutterhead are 5 r/min and 8 mm/cycle, respectively [1]. Then, the crack load excitation time cycle can be calculated by T = 60/(3.7 × 5) = 3.24 s. So, the total working time $t$ is expressed by

$$t = \mathrm{N} \times \mathrm{T} = 2.7001 \times 10^6 \times 3.240 = 8.748 \times 10^6 \ \mathrm{s} \tag{9}$$

So far, the shield machine tunneling speed $v$ is

$$v = w \times p = 40 \ \mathrm{mm/min} \tag{10}$$

Therefore, the fatigue life of the stable crack propagation stage is converted to the tunneling mileage $L$ as follows:

$$L = v \times t = 8.748 \times 10^6 \times 40 \div 60 = 5.832 \text{ km} \tag{11}$$

It can be concluded that this type of cutterhead will break completely when the tunneling mileage is 5832 km. In this subway tunnel project, the tunneling mileage of the metro section is about 2 km. It is assumed that the structure of the cutterhead may enter the stage of crack propagation directly due to welding defects. According to the above calculation, it can be concluded that the cutterhead will have cracks in the process of tunneling, and the crack depth of the cutterhead is about 17.054 mm. The depth of the crack is consistent with the actual situation. The appearance of fatigue crack on the cutterhead will lead to the decrease in cutterhead strength, and will eventually cause fatigue crack failure of the cutterhead. Therefore, it is necessary to strengthen the flaw detection of the cutterhead where cracks easily occur. When the conventional ultrasonic flaw detection method cannot meet the requirements, a higher precision flaw detection method should be adopted.

## 5. Conclusions

The stress intensity factors at different positions and angles of cracks were analyzed using numerical failure analysis of the shield machine cutterhead. It is concluded that the crack direction is mainly driven by mode I-II compound stress intensity factors at the free ends of the crack. The angle relative error between the theoretical value of 53~60° ranges from 8.6% to 1.4%. The crack shape ratio $a/c$ between 0.3 and 0.8 in the crack propagates, and the crack fracture surface always keeps semi-elliptical shape. It is under the action of mode I–II composite stress intensity factors at the free end of the crack that the cutterhead finally produces a crack and the crack begins to expand.

The maximum allowable crack depth is calculated to be 50.23 mm through transient dynamic analysis. Criterion for crack fracture damage tolerance of a shield machine cutterhead plate is proposed. The equivalent stress intensity factor is obtained by composite criterion, and the quadratic function relationship between crack depth and stress intensity factor is fitted. This type of cutterhead will break completely when the tunneling mileage is 5832 km.

This structure of the cutterhead may have welding defects, and there is no crack initiation stage basically, and it directly enters into the crack-growth stage. The actual crack depth of the cutterhead is about 16 mm, while the calculated result is 17.054 mm. It is basically consistent with the result of the calculation, which proves the correctness of the method. It is suggested to strengthen the detection of cracks or welding defects in the dangerous position where the cutterhead is prone to crack, so as to effectively prevent the cutterhead from cracking and failure.

**Author Contributions:** J.L. conceptualization, structural design, experimental test, validation, original draft; Z.Z. modeling, review and editing, supervision; C.L., sample manufacturing, review and editing; K.S. review and editing, J.G., Data analysis, supervision. All authors have read and agreed to the published version of the manuscript.

**Funding:** This work is financially supported by the Major National Research & Development project of china (2020YFB1709502) and the Hebei Natural Science Foundation Project (E2019210275).

**Institutional Review Board Statement:** Not applicable.

**Informed Consent Statement:** Not applicable.

**Data Availability Statement:** Data sharing not applicable.

**Conflicts of Interest:** The authors declare no conflict of interest.

## References

1. Ling, J.X.; Sun, W.; Huo, J.Z.; Guo, L. Study of TBM cutterhead fatigue crack propagation life based on multi-degree of freedom coupling system dynamics. *Comput. Ind. Eng.* **2015**, *83*, 1–14. [CrossRef]
2. Sun, W.; Zhu, Y.; Huo, J.Z.; Chen, X.H. Multiple cracks failure rule for TBM cutterhead based on three-dimensional crack propagation calculation. *Eng. Fail. Anal.* **2018**, *93*, 224–240. [CrossRef]
3. Sun, W.; Ling, J.; Huo, J.; Guo, L.; Song, X. Study of TBM cutterhead fatigue damage mechanisms based on a segmented comprehensive failure criterion. *Eng. Fail. Anal.* **2015**, *58*, 64–82. [CrossRef]
4. Lu, Z.Z.; Liu, Y.M. Small time scale fatigue crack growth analysis. *Int. J. Fatigue* **2010**, *32*, 1306–1321. [CrossRef]
5. Huo, J.; Zhang, Z.; Meng, Z.; Xue, L.; Jia, G.; Chen, J. A Time-Integral Crack Propagation Model Considering Thickness Effect. *IEEE Access* **2019**, *7*, 41078–41089. [CrossRef]
6. Ye, Z.; Xang, X. The research of TBM cutterhead crack propagation rate model during full stage. *Eng. Fail. Anal.* **2020**, *118*, 104806. [CrossRef]
7. Dicecco, S.; Altenhof, W.; Hu, H.; Banting, R. High-Cycle Fatigue of High-Strength Low Alloy Steel Q345 Subjected to Immersion Corrosion for Mining Wheel Applications. *J. Mater. Eng. Perform.* **2017**, *26*, 1758–1768. [CrossRef]
8. Dong, Q.; Yang, P.; Xu, G.; Deng, J. Mechanisms and modeling of low cycle fatigue crack propagation in a pressure vessel steel Q345. *Int. J. Fatigue* **2016**, *89*, 2–10. [CrossRef]
9. Zong, L.; Shi, G.; Wang, Y.Q. Experimental investigation on fatigue crack behavior of bridge steel Q345qD base metal and butt weld. *Mater. Des.* **2015**, *66*, 196–208. [CrossRef]
10. Paris, P.; Erdogan, F. Reviews of A critical analysis of crack propagation laws. *J. Basic Eng.* **1963**, *85*, 528–534. [CrossRef]
11. Beden, S.M.; Abdullah, S.; Ariffin, A.K.; Al-Asady, N.A. Fatigue crack growth simulation of aluminium alloy under spectrum loadings. *Mater. Des.* **2010**, *31*, 3449–3456. [CrossRef]
12. Forman, R.G. Study of fatigue crack initiation from flags using fracture mechanics theory. *Eng. Fract. Mech.* **1972**, *4*, 333–345. [CrossRef]
13. Neumann, P. New experiments concerning the slip processes at propagating fatigue cracks-I. *Acta Met.* **1974**, *8*, 1155–1165. [CrossRef]
14. Beretta, S.; Carboni, M.; Cantini, S.; Ghidini, A. Application of fatigue crack growth algorithms to railway axles and comparison of two steel grades. *Proc. Inst. Mech. Eng. Part F J. Rail Rapid Transit* **2004**, *218*, 317–326. [CrossRef]
15. Zerbst, U.; Madia, M.; Beier, H.T. Fatigue strength and life determination of weldments based on fracture mechanics. *Procedia Struct. Integr.* **2017**, *7*, 407–414. [CrossRef]
16. Yu, F.; Fang, Y.; Guo, S.; Bai, R.; Yin, L.; Mansouri, I. A Simple Model for Maximum Diagonal Crack Width Estimation of Shear-Strengthened Pre-damaged Beams with CFRP Strips. *J. Build. Eng.* **2021**, *41*, 102716. [CrossRef]
17. Tanaka, Y.; Okazaki, S.; Ogawa, Y.; Endo, M.; Matsunaga, H. Fatigue limit of Ni-based superalloy 718 relative to the shear-mode crack-growth threshold: A quantitative evaluation considering the influence of crack-opening and -closing stresses-ScienceDirect. *Int. J. Fatigue* **2021**, *148*, 106228. [CrossRef]
18. Haigen, J.; Yedong, W.; Xiaomei, Y.; Kemou, X. Microstructure and fatigue crack growth behavior in welding joint of Al-Mg alloy. *Eng. Failure Anal.* **2020**, *120*, 105034. [CrossRef]
19. Li, Y.D. *Theory and Applied Fracture Mechanics*; Weapons Industry Press: Beijing, China, 2005. (In Chinese)
20. Lin, X.B.; Smith, R.A. Finite element modelling of fatigue crack growth of surface cracked plates Part II: Crack shape change. *Eng. Fract. Mech.* **1999**, *63*, 523–540. [CrossRef]

*materials*

MDPI

*Article*

# Natural Aging of Ethylene-Propylene-Diene Rubber under Actual Operation Conditions of Electrical Submersible Pump Cables

**Freddy Ignacio Rojas Rodríguez, José Roberto Moraes d'Almeida and Bojan A. Marinkovic ***

Department of Chemical and Materials Engineering, Pontifical Catholic University of Rio de Janeiro (PUC-Rio), Rio de Janeiro 22453-900, Brazil; freddyrojas@aluno.puc-rio.br (F.I.R.R.); dalmeida@puc-rio.br (J.R.M.d.)
* Correspondence: bojan@puc-rio.br; Tel.: +55-21-3527-1954

**Abstract:** Ethylene-propylene-diene monomer (EPDM) rubbers used in electric submersible pump (ESP) cables were analyzed after being aged in actual operation conditions in oil wellbores. These rubbers constitute the insulation and jacket layers of the ESP cables. EPDM rubbers from four different cables operating during different time intervals (2 and 4.8 years) at different depths (from 760 to 2170 m) below sea level were studied. To verify the effects of the long exposure on the rubber performance, thermal analysis was performed to determine the thermal stability and activation energy of degradation. In addition, structural analysis, through vibrational spectroscopy and crosslinking fraction assessment, was carried out. The mechanical properties of the aged rubbers were inferred through the measurement of hardness, while the absorption of a service fluid was studied by gravimetry. The results showed only minor changes in the thermal, structural, mechanical and barrier properties of the EPDM-based ESP cable layers. It is suggested that the thermo-oxidation mechanism followed by chain scission does not have a role in the degradation of EPDM within the aged ESP cables, and no sign of variation of crosslink fractions has been encountered. Therefore, it was concluded that EPDM-based layers seem not to be weak links in the configuration of modern ESP systems.

**Keywords:** ethylene-propylene-diene monomer; activation energy of degradation; hardness; Fourier-transform infrared spectroscopy; crosslink fraction; absorption

**Citation:** Rojas Rodríguez, F.I.; d'Almeida, J.R.M.; Marinkovic, B.A. Natural Aging of Ethylene-Propylene-Diene Rubber under Actual Operation Conditions of Electrical Submersible Pump Cables. *Materials* **2021**, *14*, 5520. https://doi.org/10.3390/ma14195520

Academic Editor: Yi Gong

Received: 18 July 2021
Accepted: 17 September 2021
Published: 24 September 2021



## 1. Introduction

Electrical Submersible Pumps (ESP) have been used in the petroleum industry for artificial oil lifting from wellbores, while ESP cables serve to supply electrical power to the pumps [1]. ESP cable configuration is complex and composed of four different layers (insulation layer, covering, jacket and metal armor) surrounding three cooper wires carrying electrical current. Two of the layers, insulation and jacket, are frequently made from EPDM rubber with mutually distinct functions and properties. The insulation layer is dedicated to envelop copper wires and, therefore, has to present a high dielectric strength [2], in addition to resistance to crude oil and gas penetration. The jacket, on the other hand, provides mechanical protection to inner layers and has to be resistant to percolation of fluids.

EPDM rubber is widely used in industrial applications due to its low electrical and thermal conductivities, high weather, ozone and polar substances resistance and low adsorption of fluids. It is used in the automotive industry as window seal [3,4], in water supplying systems as water seal rings [5], in nuclear power plants within electrical cables as insulating material [6,7] and in the oil and gas industry [8].

There are various studies on EPDM degradation, generally performed through artificial aging [5,7,9–11]. However, it is also essential to evaluate degradation of EPDM during aging under operating conditions in different industrial applications.

It is also relevant to highlight that EPDM can be used with or without fillers, depending on its specific use, although the content of EPDM-based parts is commonly a trade secret in industrial applications. Among the fillers, carbon black is frequently used in EPDM formulations, since it improves tensile strength and conductivity, although decreases elongation [12] while, at the same time, increases the resistance to UV radiation [13]. Incorporation of clays additionally increases the dielectric constant of EPDM [14].

Seo et al. [6] performed a thorough study on artificial aging of EPDM, containing more than 30 wt% of fillers, used as an insulator in electrical cables in nuclear power plants. The authors evaluated their aging degradation and lifetime after accelerated aging at temperatures between 140 and 170 °C and, also, after a loss of coolant accident (LOCA), when temperatures and pressures can reach values higher than 180 °C and 4 atm. No radiation was used for aging purposes.

The authors found [6] that temperatures higher than 140 °C are particularly harmful and can reduce the lifetime almost 10 times, as determined during aging at 170 °C. In addition, activation energies of degradation of EPDM showed reduction between 15 and 25% after LOCA conditions, in comparison to EPDM before LOCA, depending on the experimental data used for their calculation. In accordance with the authors, the reduction of activation energy was an indication of polymer degradation through mechanisms such as thermo-oxidation and chain scission. Interestingly, Fourier-Transform InfraRed spectroscopy (FTIR) did not reveal any change in EPDM spectra, before and after LOCA conditions, despite the fact that C=O and C=C bands should be expected in accordance with the degradation mechanisms and despite the significant decrease in mechanical properties.

Zhao et al. [15] studied EPDM degradation under artificial accelerated weathering conditions, using high humidity (65%) and temperature (55 °C), rain and light exposure intervals and concluded that in such severe conditions, degradation of EPDM is rapid along the first 12 days.

Šarac et al. [7] performed artificial aging of EPDM using different temperatures and radiation doses. Although this study is not directly connected to the operation conditions of EPDM in ESP cables, it is still a rare study of artificial EPDM aging. They reported that at room temperature, ultimate tensile stress increased firstly with an increase in absorbed radiation doses until 600 kGy and decreased for the higher doses. Such a behavior of the ultimate tensile stress has been explained as the interplay between crosslinking, predominant at lower doses, with the chain scission mechanism, dominating at higher radiation doses.

Wang et al. [9] reported that the tensile strength and elongation at break decreased with time of artificial aging at seawater conditions, while hardness increased. The authors pointed to chain scission as the mechanism causing degradation of EPDM under seawater conditions.

Nakamura et al. [5] reported a rare study of natural aging of EPDM under actual operation conditions. They investigated degradation of EPDM used in seal rings in drinking water supplying systems, since degradation fragments might contaminate water. These EPDM seals have been naturally aged for 3 years in contact with the water supplied in Osaka city, Japan, at temperatures varying between 20 and 45 °C. The authors encountered a decrease in hardness of about 10% and almost 40% reduction in crosslinking density in comparison to unused EPDM seals. Scanning Electron Microscopy (SEM) evidenced a significant degradation of naturally aged EPDM seals through observation, since degradation fragments might contaminate water. Degradation mechanisms were attributed to thermo-oxidation of EPDM and chain scission. FTIR spectrum of aged seals exhibited vibrational bands at 1640 and 1720 $cm^{-1}$ identified as C=C and C=O bonds, respectively, while in the FTIR spectrum of unused seals, these bands were absent. It is worth noting that chain scission is responsible for the formation of C=C bonds inside EPDM.

Li et al. [16] studied thermo-oxidative and compressive stress-thermo-oxidative aging of EPDM and concluded that activation energy decreases with the increase in compressive stress. This indicated that compressive stress contributes to accelerate degradation mechanisms in EPDM.

Awwad et al. [11] evaluated the aging of EPDM exposed to a NaOH solution (25 wt% NaOH) at 38, 54 and 77 °C, as well as to water at 77 °C for 12 months. Tensile strength decreased as a function of temperature and exposure time. The longer the time and the higher the temperature, the greater the decrease in mechanical properties. This behavior was attributed to the NaOH solution penetration inside EPDM samples.

ESP cables are currently exchanged in regular time spans to avoid cable deterioration and, consequent, non-planned interruption of oil production. Despite their importance in the petroleum industry, there is a complete lack of knowledge on natural or artificial aging of EPDM components, used for fabrication of insulation and jacket layers, inside ESP cables, as far as the authors are aware of. In addition, ESP recently started to employ permanent magnet motors, increasing significantly their service life time [1]. Therefore, the question has been raised if ESP cables become weak links inside the ESP system.

In order to elucidate changes of physico-chemical properties of EPDM used for fabrication of insulation and jacket layers, ESP cables naturally aged inside oil wellbores in actual operation conditions over 738 (~2 years) and 1752 days (~4.8 years) at different depths, with respect to sea level, were studied and compared with a virgin ESP cable.

The principal goal of this research was to evaluate the extent of degradation of insulation and jacket EPDM-based layers during actual operation conditions of ESP cables by studying their hardness, thermal stability, activation energy of degradation, chemical structure (through vibrational spectroscopy and crosslink fraction) and barrier capacity. The filler composition was also elucidated for both layers.

This work presents, as a novelty, the natural aging study of EPDM layers (insulation and jacket) within ESP cables after their use under actual service conditions and intends to evaluate whether these cables can become the weak link in new ESP systems, in which pumps with much greater durability are being used. This knowledge can be used as a guide for avoiding unscheduled interruptions of oil production and, therefore, for minimization of operating costs.

## 2. Experimental

### 2.1. Materials

Five ESP cables were kindly supplied by Equinor Brasil Energia Ltd. (Rio de Janeiro, Brazil). Two ESP cables (denoted, A and B) spent 738 days (~2 years) in real operation conditions at the depths of 917 and 1927 m, respectively. Two other ESP cables (denoted, C and D) spent 1752 days (~4.8 years) in real operation conditions at the depths of 760 and 2170 m, respectively. The temperatures of ocean water at the depths between 760 and 2170 m are approximately between 8 and 4 °C, while the pH of liquid in contact with ESP cables was ~8.

It is relevant to notice that the ESP cables A and B operated in a different wellbore from that of the cables C and D. The fifth cable, denoted E, was a virgin ESP cable. The metallic armor was made from galvanized steel for all studied ESP cables. A schematic draw of ESP cable configuration, with its principal layers, is presented in the Supporting Information, Figure S1.

In accordance with the datasheet of the two EPDM-based layers, insulation and jacket, these were made from EPDM DL90 and EPDM CL-185, respectively.

### 2.2. Characterization Techniques

Fourier Transform InfraRed spectroscopy (FTIR) was performed on a PerkinElmer Spectrum Two FT-IR-ATR spectrometer. Small quantities of insulation and jacket layers, from five ESP cables, were used for the analyses without any additional treatment. Spectra were recorded in the range from 400 to 4000 $cm^{-1}$, with a resolution of 4 $cm^{-1}$ and 20 scans.

Thermogravimetric analyses (TGA) were performed using a PerkinElmer Simultaneous Thermal Analyzer, model STA 6000 (PelkinElmer, Beaconsfield, England). Two different experimental conditions were used to analyze EPDM samples, regardless of the layer.

Under **condition I**, 10–15 mg of EPDM were heated under $N_2$ flow (20 mL $min^{-1}$), at heating rates of 5, 10, 15, 20, 25, 30 and 40 °C $min^{-1}$ in the temperature range of 30–570 °C. Therefore, 7 runs were carried out per EPDM from each cable and from each EPDM layer (insulation and jacket). This totalized 70 runs carried out under **condition I**.

In the experiments using **condition II**, 10–15 mg of EPDM were heated from 30 to 570 °C under $N_2$ flow (20 mL $min^{-1}$), at a heating rate of 25 °C $min^{-1}$, while from 570 to 930 °C gas flow was changed to synthetic air (20 mL $min^{-1}$), while the heating rate was kept the same. This totalized 10 experiments, considering both EPDM layers, performed under **condition II**.

Crosslink fractions of EPDM from insulation and jacket layers were calculated following the experimental and data analyses methods described in ASTM D 2765-01 [17,18]. It is relevant to mention that this ASTM standard considers EPDM exclusively part of the total EPDM formulation (consisting of EPDM matrix and organic and inorganic fillers and additives) for calculation of crosslink fractions. At least three measurements were performed per EPDM sample.

The Shore D hardness test was performed with a Digimess durometer, based on the ASTM D 2240-15 standard.

For the insulation layer, cylindrical specimens were prepared and measurements were taken along the length of the specimen, as well as on the cross section, along the perimeter. For the measurement of hardness in the cross section, the specimens were only 50 mm long, in order to avoid any buckling effect during the measurement. On the other hand, the measurements along the length were made on 110 mm long specimens. The copper wires were not removed from the cylindrical specimens to avoid damaging the insulation EPDM layer. Nine measurements were taken along the length of the samples and eight measurements along the perimeter. Three samples were used per cable. Therefore, each data point (hardness of each EPDM specimen) consists of an average of 27 measurements for the hardness measured along the length of each cable, and 24 measurements for the hardness measured along the perimeter of each cable.

For the jacket layer, tests were performed on the cross section of cylindrical specimens cut from ESP cables. Four specimens were machined from each cable and measurements were carried out on both sides of the samples (top and bottom faces). On each face, 15 measurements were made, totalizing 120 measurements per cable.

Absorption tests were performed on the insulation and jacket rubber layers. The tests were based on ASTM D 570-98 standard, with the aim to determine the amount of fluid absorbed by gravimetry. For both layers, three samples from each cable—A to E—were immersed in the service Safe-Scan + Packer fluid (2.22% $v/v$) for up to 3777 h (~158 days). This is a protection fluid for ESP cables and was prepared with the formulation commonly used when the cables are in regular operation conditions. The samples of each cable were immersed in different containers to avoid any type of contamination/interference from samples coming from different cables. The samples were removed periodically from the containers and weighted within ±0.001 g, to determine the mass gain with respect to the immersion time.

The samples of the insulation layer, consisted of cylinders 60 mm long, from which the copper wires, that are surrounded by the insulation layer, have not been removed. This procedure was adopted considering that the removal of the copper wire produced too much damage in the rubber layer. Furthermore, any absorption of the fluid by the copper wire is not expected. Thus, the presence of the wire does not affect the measurement procedure. For the jacket layer, cylindrical specimens were used as well but 50 mm long. These specimens consisted of rubber segments, from which the external metallic armor and the insulation layer/copper wires assembly was removed.

## 3. Results

FTIR spectroscopy permits the identification of the thermo-oxidative degradation mechanism in EPDM during aging and also allows the identification of additives incorporated in EPDM formulations within the insulation and jacket layers.

Figure 1a–e shows FTIR spectra of the insulation layer from the ESP cables A to E. The vibrational bands attributed to EPDM [6,19,20] are listed in Table 1. In addition to EPDM vibrational bands, all FTIR spectra exhibit the same group of additional bands attributed to metakaolinite [21], a product of thermal decomposition of kaolinite clay mineral at temperatures > 530 °C. It is worth noting also that the weak bands situated at 1720 and 1640 $cm^{-1}$, attributed to carbonyl (C=O) and carbon–carbon double bonds (C=C) [5], respectively, were identified in the spectrum of the virgin ESP cable (Figure 1e), and also in the spectra of the naturally aged ESP cables during 738 and 1752 days at different depths (Figure 1a–d). The intensities of these two vibrational bands, which are the fingerprints of oxidation and chain scission degradation mechanisms of EPDM [5,6,22], apparently keep similar values in the virgin and naturally aged cables.

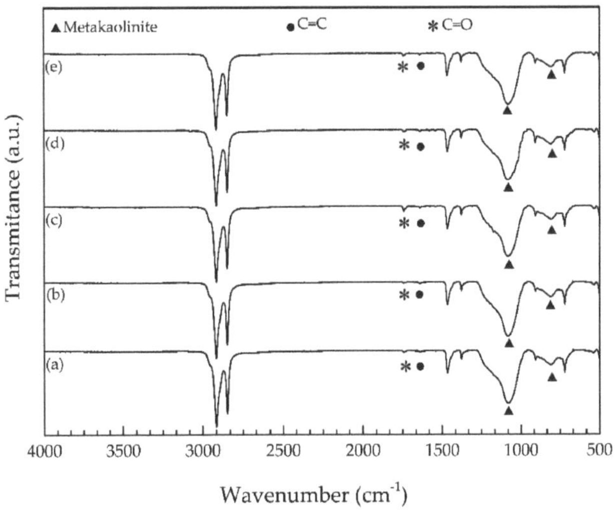


**Figure 1.** FTIR spectra of the insulation layer of cables (**a**) A; (**b**) B; (**c**) C; (**d**) D and (**e**) E. ▲ stands for metakaolinite, • stands for C=C and * stands for C=O bands.

FTIR spectra of EPDM composing a jacket layer of the ESP cables A to E are presented in Figure 2a–e. All spectra showed vibrational bands belonging to EPDM [6,19,20], as listed in Table 1. Furthermore, all spectra present additional bands similarly to that previously noted for the insulation layer. A peculiarity is that the EPDM from the cables A and B (aged for 738 days at the depths of 917 and 1927 m, respectively) presented vibrational bands of kaolinite [21], while EPDM from the cables C, D (aged for 1752 days at the depths of 760 and 2170 m, respectively) and E (virgin cable) presented vibrational bands of metakaolite [21]. As in the case of EPDM from the insulation layer, FTIR spectra of EPDM from the jacket also presented vibrational bands with similar intensities originated from carbonyl and carbon–carbon double bounds, regardless of whether the cable was virgin or naturally aged for 738 and 1752 days.

**Table 1.** Vibrational bands of EPDM.

| Band ($cm^{-1}$) | Interpretation | Description |
|---|---|---|
| 3075 | $\upsilon_{sy}SP^2 = C - H$ | Symmetrical stretch C–H |
| 2957 | $\upsilon_{asy}SP^3\ C - H$ | Asymmetric stretch C–H |
| 2915 | $\upsilon_{asy}SP^3\ C - H_3$<br>$\upsilon_{asy}SP^3\ C - H_2$ | Asymmetric stretch C–H |
| 2848 | $\upsilon_{sy}SP^3\ C - H_3$<br>$\upsilon_{sy}SP^3\ C - H_2$ | Symmetrical stretch C–H |
| 1455 | $\delta_{asy}SP^3\ C - H_3$<br>$\delta_{sy}SP^3\ C - H_2$ | Asymmetric angular deformation in the plane $CH_3$Symmetrical angular deformation in plane (scissor) $CH_2$ |
| 1374 | $\delta_{sy}SP^3\ C - H_3$ | Symmetrical angular deformation in the plane (umbrella) $CH_3$ |
| 709 | $\delta_{\rho}SP^3\ CH_2$ | Asymmetric angular deformation off plane (wagging) $n_{CH_2} \geq 4$ |

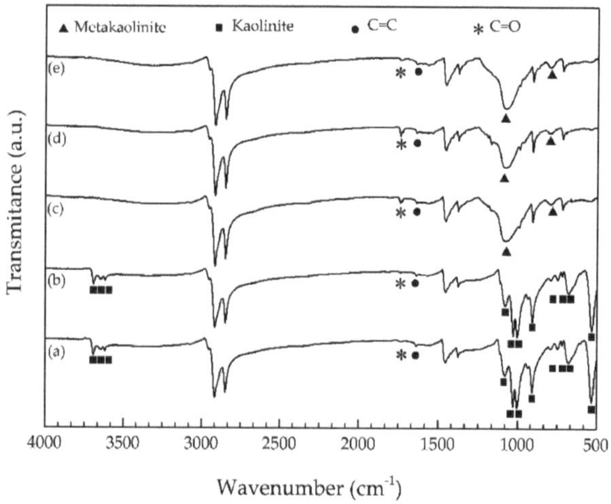


**Figure 2.** FTIR spectra of jacket layer of cables (**a**) A; (**b**) B; (**c**) C; (**d**) D and (**e**) E. ■ stands for kaolinite, • stands for C=C and * stands for C=O bands, ▲ stands for metakaolinite.

TGA has been conducted to identify degradation stages, to quantify content of additives, organic and inorganic, presented in EPDM formulation and, mainly, to calculate activation energy of degradation of EPDM, in order to shed more light on EPDM degradation, which might occur during natural aging.

Figure 3 shows TGA curves of the insulation layer of the virgin ESP cable and for naturally aged cables under condition I. The weight loss stages and contents are very similar for all cables, virgin or aged (Table 2). Three stages of weight loss could be determined, at most, although in the first two, up to 410 °C, the weight loss is only slightly higher than 3 wt%. The small loss at the first stage until 250 °C (<1 wt%) may be attributed to humidity and some small organic molecules. The second stage loss, from 250 to 410 °C, is commonly ascribed to evaporation of extender oils, generally used in fabrication of EPDM to reduce viscosity and as fillers. In the third stage, from 410 to 520 °C, a weight loss, on average, as high as 59 wt% was recorded, a consequence of decomposition and volatilization of EPDM [6]. On average, about 37.5 wt% of the insulation layer remained after heating to 570 °C.

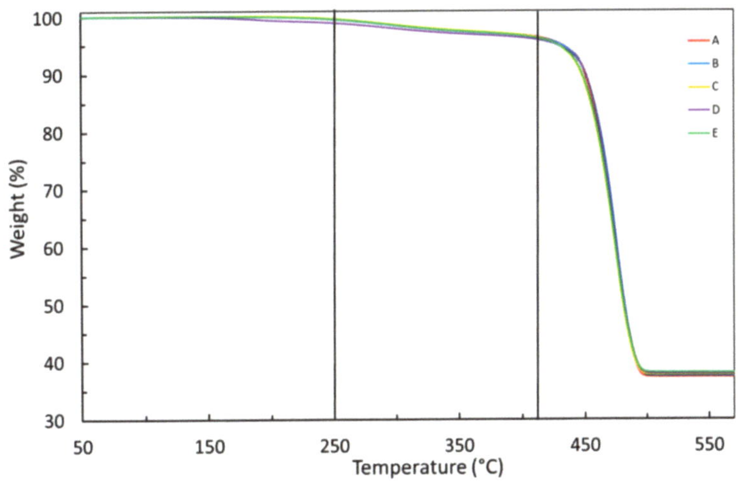


**Figure 3.** TGA curves of the insulation layer for virgin and naturally aged cables, under condition I (heating rate of 10 °C min$^{-1}$). A–E stand for different ESP cables as denoted in the Section 2.

**Table 2.** Weight losses of the insulation layer for the cables A–E evaluated under condition I.

| Cable | %Weight Loss 30–250 °C | %Weight Loss 250–410 °C | %Weight Loss 410–570 °C |
|---|---|---|---|
| A | 0.3 | 2.9 | 59.7 |
| B | 0.3 | 2.9 | 59.8 |
| C | 0.3 | 3 | 59.7 |
| D | 0.9 | 3.3 | 58.7 |
| E | 0.3 | 2.8 | 59.4 |
| **Average** | **0.4** | **3** | **59.5** |
| **Standard deviation** | **0.2** | **0.2** | **0.4** |

Under condition II, carried out until 930 °C, insulation layers of all five ESP cables showed the same mass loss stages and loss contents (Figure 4) as previously observed under condition I, indicating that the remaining mass of about 37.5 wt% is due to an inorganic filler such as metakaolinite, previously identified by FTIR (Figure 1).

Interestingly, activation energies of degradation ($E_a$) of insulation layers, calculated by the Kissinger model [23,24] for virgin and naturally aged cables are distributed within a few percentages around the value of the virgin cable of ~233 kJ mol$^{-1}$ (Figure 5 and Table 3).

EPDM from the jacket layer also demonstrated three stages of weight loss under the **condition I** until 520 °C, while from 520 to 570 °C, no weight loss was detected (Figure 6). In the first stage (from 30 to 250 °C) the weight loss is low (Table 4), on average slightly higher than 1 wt%, and was attributed to water loss and, possibly, to the loss of some other low-weight organic molecule. During the second stage, from 250 to 410 °C, a weight loss of on average 14 wt% (Table 4) was detected and attributed to the loss of extender oils. In the last stage, from 410 to 520 °C, on average 56 wt% were lost (Table 4), which is in accordance with the well-known temperature range of EPDM decomposition [6]. Therefore, on average, 29 wt% of an inorganic filler remained present, and at least a portion of it can be ascribed to metakaolinite or kaolinite, depending on the cable, as pointed out by FTIR analyses (Figure 2).

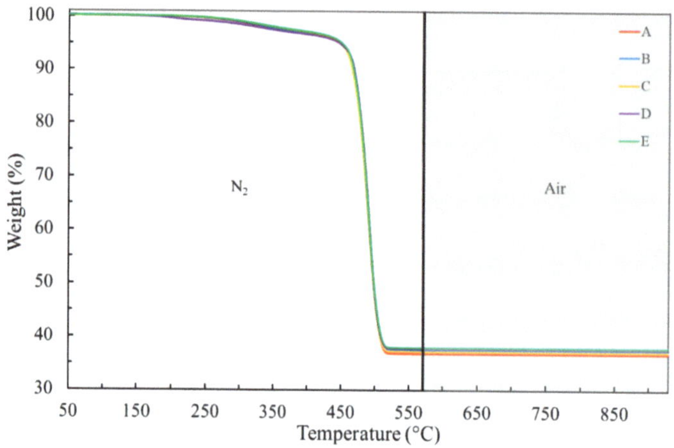


**Figure 4.** TGA curves of the insulation layer for virgin and naturally aged cables, under condition II (heating rate of 25 °C min$^{-1}$). A–E stand for different ESP cables as denoted in the Section 2.

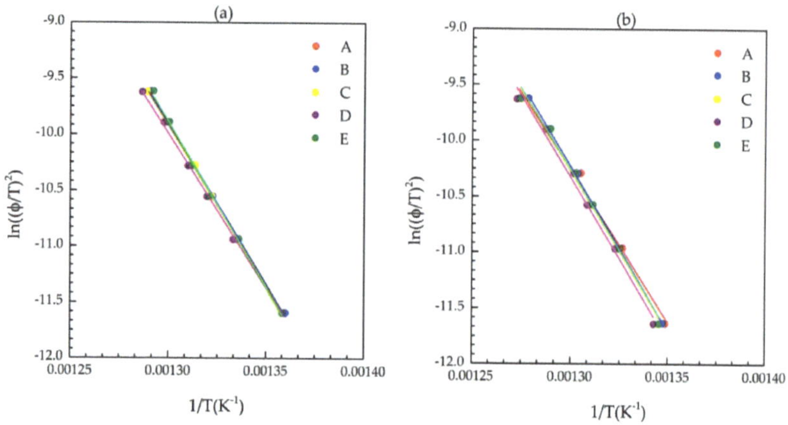


**Figure 5.** Kissinger plots (**a**) insulation layer (**b**) jacket layer. A–E stand for different ESP cables as denoted in the Section 2.

**Table 3.** Activation energies of the insulation and jacket layers, as calculated by the Kissinger method.

| | $E_a$ (kJ mol$^{-1}$) | |
|---|---|---|
| **Cable** | **Insulation Layer** | **Jacket Layer** |
| A | 227.20 | 227.20 |
| B | 242.00 | 217.89 |
| C | 235.48 | 217.96 |
| D | 235.49 | 216.94 |
| E | 232.93 | 237.83 |

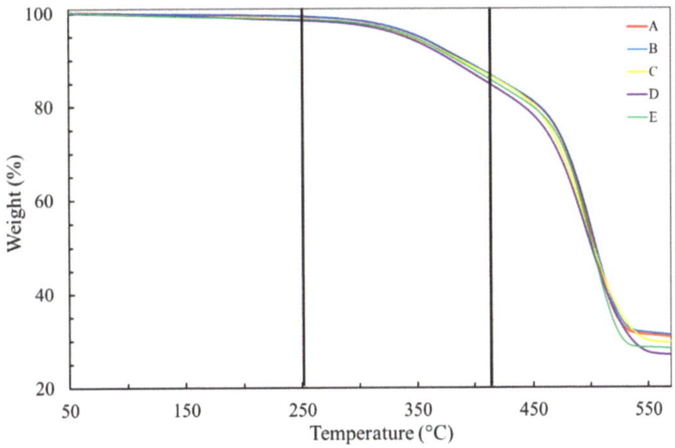


**Figure 6.** TGA curves of jacket layers for virgin and naturally aged cables, under condition I (heating rate of 25 °C $min^{-1}$). A–E stand for different ESP cables as denoted in the Experimental section.

**Table 4.** Weight losses of the jacket layer for the cables A–E evaluated under condition I.

| Cable | %Weight Loss 30–250 °C | %Weight Loss 250–410 °C | %Weight Loss 410–570 °C |
|---|---|---|---|
| A | 0.7 | 13.6 | 54.6 |
| B | 0.6 | 13.8 | 54.1 |
| C | 1.4 | 13.1 | 55.8 |
| D | 1.6 | 14.7 | 56.3 |
| E | 1.4 | 14.1 | 55.8 |
| **Average** | **1.2** | **13.9** | **55.3** |
| **Standard deviation** | **0.4** | **0.5** | **0.8** |

In order to have a better understanding of filler types inside EPDM, composing about 29 wt% of the total EPDM formulation in the jacket layer, the **condition II** was applied. Figure 7 demonstrates that the switch of atmosphere at 570 °C, from inert ($N_2$) to oxidative (synthetic air), is capable of identifying the presence of carbon black through the weight loss, between 700 and 800 °C, probably emitted as $CO_2$. The weight lost due to carbon black is responsible for, on average, 18 wt%, while 8.5 to 14 wt% of residual material, depending on particular ESP cable, still presented at 930 °C, were due to metakaolinite and kaolinite fillers (Table 5).

Kissinger plots for EPDM from the jacket layer for all cables are shown in Figure 5 and the as-calculated values for the energy of activation are summarized in Table 3. It should be highlighted that the activation energies were calculated from seven experimental points (as it was the case for the insulating layer), offering a robust amount of data for their calculation.

The value for the energy of activation for degradation of EPDM, in jacket layer, for the virgin cable (E) is 237.83 kJ $mol^{-1}$ and is higher than those calculated for the naturally aged cables (Table 3). EPDM naturally aged inside the cables B, C and D showed reduction of activation energy for their degradation of about 10% in comparison to the activation energy of virgin cable, while this reduction was about 5% for the EPDM from the cable A, which has been exposed to milder conditions in terms of time (738 days) and depth (917 m), in comparison to the other aged cables (B, C and D).

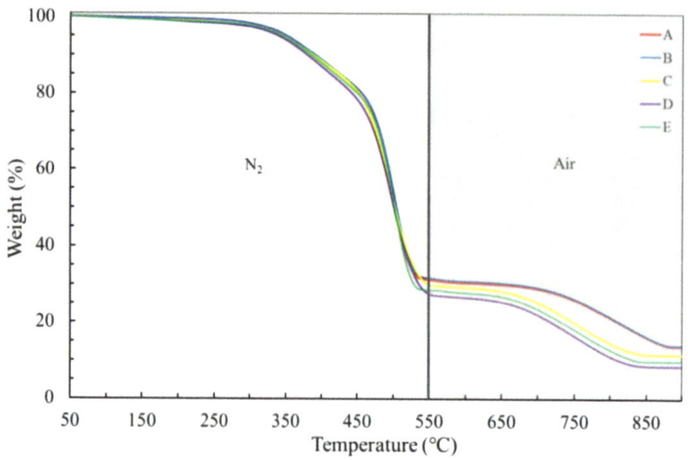


**Figure 7.** TGA curves of the jacket layer for virgin and naturally aged cables, under condition II (heating rate of 25 °C min$^{-1}$). A–E stand for different ESP cables as denoted in the Experimental section.

**Table 5.** Weight losses and residues for the cables A–E, evaluated under condition II.

| | % of Weight Loss from 570 to 930 °C (Carbon Black) | | | | |
|---|---|---|---|---|---|
| **Cable** | A | B | C | D | E |
| | 17.06 | 17.35 | 18.48 | 18.57 | 18.46 |
| | **% of Weight Residue at 930 °C (A, B: Kaolinite and C, D, E: Metakaolinite)** | | | | |
| **Cable** | A | B | C | D | E |
| | 13.86 | 14.07 | 11.61 | 8.59 | 9.89 |

The results of Shore D hardness tests on insulation layers of cables A to E show no significant variation in hardness as a function of service time or depth of operation, both in longitudinal and transversal modes (Figure 8a,b).

Comparing longitudinal vs. transversal hardness, there is a tendency for the values measured longitudinally to be slightly higher than those measured transversely. In addition to these small differences, it is reasonable to expect that both hardness values may deviate from the absolute values due to the shape/dimensions of the specimens.

The longitudinal measurements may present slightly overestimated values due to the small thickness of the rubber layer over the copper wire. Thus, it could be argued that the measured hardness value might be higher than expected due to the contribution of the copper substrate. On the other hand, the transverse hardness, also due to the small thickness of the rubber layer, may not be sufficiently rigid to confine the hardness dent and thus resulted in an underestimated hardness value.

If any of these experimental deviations are occurring, they are small and not significant for the general analysis, as can be seen from the results shown in Figure 8a,b.

The results of the Shore D hardness test of the EPDM jacket layer are shown in Figure 8c. Interestingly, their hardness was significantly lower (>20%) in comparison to the hardness of the insulating layer. However, similarly as found for the EPDM insulation layer, there was no significant variation in the hardness between the EPDM rubber of virgin and naturally aged cables.

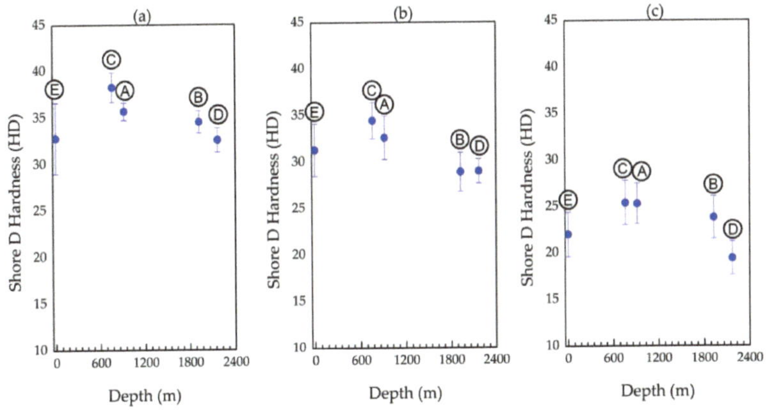


**Figure 8.** Hardness (**a**) along a longitudinal section of the insulation layer, (**b**) along a transversal section of the insulation layer, and (**c**) of the jacket layer. A–E stand for different ESP cables as denoted in the Experimental section.

Table 6 summarizes crosslink fractions of EPDM of the insulation layers, together with their standard deviations. The crosslink fractions of the virgin and naturally aged cables vary within the standard deviation. A maximum variation of 5.7% for the isolation layer and of 5.8% for the jacket layer were obtained.

**Table 6.** Crosslink fractions for insulation and jacket layers.

| Cable | Insulation Layer | Standard Deviation | Jacket Layer | Standard Deviation |
|---|---|---|---|---|
| A | 0.91 | 0.005 | 0.65 | 0.03 |
| B | 0.92 | 0.02 | 0.72 | 0.02 |
| C | 0.91 | 0.03 | 0.72 | 0.04 |
| D | 0.84 | 0.02 | 0.73 | 0.006 |
| E | 0.87 | 0.02 | 0.69 | 0.02 |

The crosslink fractions of the jacket layers (Table 6) are significantly lower than the fractions calculated for the EPDM from the insulating layers. This is in accordance with their lower hardness, as previously measured by Shore D (Figure 8a–c). The crosslink fractions within the jacket vary in a narrow range close to 0.70, pointing out that this lower fraction interval is a characteristic of the EPDM used for fabrication of the jacket layer. No indication of the EPDM degradation in the jacket layer of naturally aged cables was evidenced by reduction, or increase, of crosslink content.

Figure 9 shows the results of the absorption test for all analyzed samples. It can be observed that the insulation layer absorbed insignificant amounts of the fluid. The small variation found (±0.01%) was considered to be within the experimental error of the test, mainly due to the process of excess fluid removal from the sample surface at each weighing.

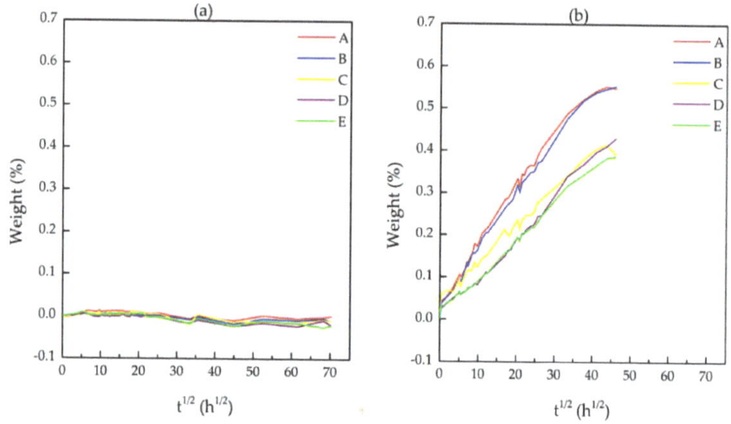


**Figure 9.** Adsorption plots of (**a**) insulation layer and (**b**) jacket layer. A–E stand for different ESP cables as denoted in the Experimental section.

The results obtained for the jacket layer indicate a small absorption of up to about 0.55% after 158 days of immersing the samples in the Safe-Scan + Packer fluid. As can be seen, the absorption process did not reach complete saturation and there was a significant difference between the rubbers of cables A and B and C and D. Cables C and D present the same behavior found for the virgin cable (cable E). It should be remembered here that the cables A and B come from a different wellbore from that of the cables C and D. It is also important to note that the operating time of cables A and B (738 days) was much shorter than that of cables C and D (1752 days); however, there was greater absorption in cables A and B.

## 4. Discussion

Based on the intensities of the vibrational bands located at 1720 and 1640 cm$^{-1}$, belonging to C=O and C=C vibrational bands, respectively, FTIR spectra of insulation and jacket layers (Figures 1 and 2) did not support progress of thermo-oxidation and chain scission mechanisms of degradation during natural aging of ESP cables. Carbonyl groups and C=C bonds have already been introduced, in small amounts, during the fabrication process of EPDM formulations for both insulation and jacket layers, and FTIR was not capable of discerning further degradation of EPDM through these mechanisms during natural aging. Similarly, Seo et al. [6] did not find evidence from FTIR analyses on oxidation degradation after LOCA aging, through the appearance of the C=O band, although aged EPDM demonstrated significant reduction of mechanical properties, life time and even a decrease in activation energy of degradation at the level as high as 25%.

However, Nakamura et al. [5] were capable of observing the appearance of C=O and C=C vibrational bands in seal rings aged during 3 years in actual operation conditions, including temperatures as high as 45 °C. Li et al. [16] evidenced that the aging temperature is important for C=O formation on thermal-oxidation. They observed that the C=O band started to increase, in comparison to the unaged EPDM, only when aging temperature reached 70 °C. The same authors identified low intensity C=O vibrational bands also on unaged EPDM, corroborating the observation that some low extent of thermo-oxidation may already occur during fabrication. Wang et al. [25] also observed molecular chain degradation, bond breaking and EPDM oxidation, but they conducted their experiments at high temperatures (130, 145, and 160 °C).

Therefore, the absence of further oxidation in the EPDM inside insulation and jacket layers can be understood in terms of low temperatures of sea water (although not in direct contact with ESP cables), as reported in the Experimental section, and the lack of oxidation atmosphere at the aging depths, the conditions that do not promote thermo-oxidation.

In addition to the FTIR results, it is relevant that the activation energies of degradation for the aged insulation layer (Table 3), as calculated by TGA data through Kissinger analysis, are very close to the activation energy of the virgin cable (maximum variation of 3.9%), indicating the absence of degradation. However, activation energies of degradation for the jacket layer (Table 3) of aged cables showed a small decrease, suggesting that a degradation mechanism has been active, especially for the more severely aged cables B, C and D, which presented a reduction in activation energy as high as 10%. In comparison to the reduction of about 25% after LOCA aging reported by Seo et al. [6], the reductions of activation energy presented by EPDM from the jacket layer were moderate, situated between 10% (for more severe aging conditions in terms of time and depth) and 5% (for the least severe condition). The Ozawa–Flynn–Wall model, together with three other models, Friedman, combined kinetic analysis and correction of the combined kinetic analysis by isoconversion, were also used to estimate activation energy of EPDM degradation in the jacket layer. These results, to be present elsewhere, showed a similar trend of small or minor changes of activation energies with natural aging conditions.

The small decrease in activation energy of EPDM in the jacket layer, apparently, 5% to 10% as cited above, is not related to any kind of oxidation degradation as corroborated by FTIR but is probably connected to chemical degradation or hydrolysis induced by fluids and humidity, respectively, as inferred from the work of Woo et al. [26]. It is worth noting that the jacket layer is closer to the surface of the ESP cable and, therefore, is more susceptible to interaction with the chemical agents from the environment.

The higher hardness of EPDM in the insulation layer in comparison to the jacket (Figure 8) is due to higher, about 20%, crosslink fraction of EPDM inside the insulation layer (Table 6). Furthermore, EPDM formulation for the insulation layer is richer in inorganic content than for the jacket layer (Tables 2, 4 and 5), which can additionally contribute to the higher hardness of EPDM in the insulation layer. The hardness of EPDM in the insulation layer is at the same level of Shore D scale reported by Kömmling et al. [27]. In addition, these authors found that the hardness of EPDM hardly changes with aging for temperatures $\leq$ 125 °C, while at the higher temperatures of aging (>125 °C), hardness increased several times, owing to additional crosslinking [28,29]. In accordance with the findings of Kömmling et al. [27], hardness of EPDM from insulation and jacket layers also hardly changed with natural aging, as observed from Figure 8.

In accordance with hardness measurements, crosslink fractions (Table 6) did not change significantly for the insulation or jacket layer. This would seem to show that EPDM in both layers did not suffer from a significant change in reticulation, as experimentally observed.

The higher absorption of jacket layer is attributed to fewer reticulated EPDM chains in this layer when compared to EPDM in the insulation layer (much more crosslinked). However, both EPDM layers absorbed an insignificant amount of Safe-Scan + Packer fluid, while it is worth noting that neither of the layers is actually in direct contact with environment liquids during service life time. Thus, both EPDM-based layers kept high barrier properties during aging.

The absorption tests (Figure 9) also seem to indicate an influence of the wellbore chemical environmental conditions on the absorption process. In fact, the interaction of EPDM rubber with organic compounds will vary depending on the type of compound, as expected but also with the volume fraction of different compounds in a mixture. For example, the mass uptake of aromatic hydrocarbons is higher than that of alkanes in EPDM rubber, and mixtures of these compounds increase the effect of absorption when the temperature is increased [30]. Since a wellbore contains a mixture of numerous hydrocarbons, deleterious synergistic effects can occur, which seems to be the case for the wellbore where cables A and B operated. Moreover, the deleterious effect of fluid absorption can be enhanced with the presence of $CO_2$ in solution [31], and $CO_2$ is one of the three most common gases found in oil wellbores, along with $H_2S$ and $CH_4$ [32].

The present data strongly suggested that EPDM in insulation and jacket layers of ESP did not suffer, during prolongated aging, from thermo-oxidation degradation, which is detectable through the formation of the carbonyl group and chain scission. In addition, no sign of crosslink fraction change has been found for the studied EPDM layers even after being in operation for time periods as long as 4.8 years. It is worth noting that only EPDM from the jacket layer presented an indication of slight degradation through a low decrease in activation energy (10% at most). It can be speculated at this point that this reduction in activation energy is related to another type of aging, such as chemical degradation or hydrolysis, as pointed out by Woo et al. [26] in their study.

The most important findings of our study, although taken on a limited number of specimens, are that EPDM-based layers from ESP cable seem not to be weak links within the modern ESP systems with longer service life, since no significant degradation was found even after aging in actual service conditions for the periods as long as 4.8 years and at depths deeper than 2000 m.

## 5. Conclusions

This study found a small effect of aging, in actual operation conditions of ESP cables, on the mechanical, thermal, structural and barrier properties of the two EPDM-based layers (insulation and jacket).

It is suggested that the thermo-oxidation mechanism followed by chain scission does not have a role in the degradation of EPDM within the aged ESP cables, and no sign of increase or decrease in crosslink fraction has been encountered.

Hardness of EPDM hardly changed in both layers during natural aging in actual operation conditions of ESP cables.

While EPDM from the insulation layer did not present changes in barrier properties on aging, EPDM from the jacket layer of the cables A and B (2 years of aging at 917 m and 1927 m, respectively) showed slightly lower barrier properties than the EPDM from the cables C and D (4.8 years of aging at 760 and 2170 m, respectively) and from the virgin cable E. This difference was attributed to possibly different chemical environments in two different wellbores, since the cables A and B served in a different wellbore from the cables C and D.

Activation energies of degradation calculated by the Kissinger method indicated a modest decrease (up to 10%) only for the EPDM from jacket layers.

The previous data pointed out that slight degradation of EPDM from the jacket layer, detected mainly through a reduction of activation energy, is due to some non-oxidation mechanism, such as, for example, chemical degradation and/or hydrolysis. To complement the data obtained in this work, it is intended to perform nuclear magnetic resonance spectroscopy of virgin and aged EPDM samples.

It can be concluded from our study that EPDM-based layers from ESP cables seem not to be weak links in the structure of modern ESP systems, although a higher number of ESP cables from different wellbores, naturally aged in actual operation conditions during periods of several years, and at different depths, would be a mandatory requirement to acquire a more robust data set and knowledge on degradation processes inside these cables. In addition, there are no current acceptance criteria for the lifespan of ESP cable parts and, therefore, these need to be established.

**Supplementary Materials:** The following are available online at https://www.mdpi.com/article/10.3390/ma14195520/s1, Figure S1: Schematic draw of ESP cable configuration, with its principle layers.

**Author Contributions:** Conceptualization, J.R.M.d. and B.A.M.; Overall characterizations, F.I.R.R.; Funding acquisition, B.A.M.; Methodology, F.I.R.R., J.R.M.d. and B.A.M.; Data interpretation, F.I.R.R., J.R.M.d. and B.A.M.; Writing—original draft, B.A.M.; Writing—review & editing, F.I.R.R., J.R.M.d. and B.A.M. All authors have read and agreed to the published version of the manuscript.

**Funding:** This study was financed by Equinor Brasil Energia Ltd.a through the contract 21123-3.

**Institutional Review Board Statement:** Not applicable.

**Informed Consent Statement:** Not applicable.

**Data Availability Statement:** The data underlying this article will be shared on reasonable request from the corresponding author.

**Conflicts of Interest:** The authors declare no conflict of interest.

## References


1. Takacs, G. *Electrical Submersible Pumps Manual*; Elsevier: Burlington, NJ, USA, 2009; Volume 1, ISBN 9781856175579.
2. Su, J.; Zhang, J. Reinforced properties of ethylene–propylene–diene monomer composites by vinyltrimethoxysiloxane functionalised barium titanate. *Plast. Rubber Compos.* **2016**, *45*, 127–135. [CrossRef]
3. Yang, Z.; Peng, H.; Wang, W.; Liu, T. Crystallization behavior of poly(ε-caprolactone)/layered double hydroxide nanocomposites. *J. Appl. Polym. Sci.* **2010**, *116*, 2962–2972. [CrossRef]
4. Jacob, C.; Bhowmick, A.K.; De, P.P.; De, S.K. Studies on ground EPDM vulcanisate as filler in window seal formulation. *Plast. Rubber Compos.* **2002**, *31*, 212–219. [CrossRef]
5. Nakamura, T.; Chaikumpollert, O.; Yamamoto, Y.; Ohtake, Y.; Kawahara, S. Degradation of EPDM seal used for water supplying system. *Polym. Degrad. Stab.* **2011**, *96*, 1236–1241. [CrossRef]
6. Seo, Y.-D.; Lee, H.-S.; Kim, Y.-S.; Song, C. A Study on the Aging degradation of ethylene-propylene-diene monomer (EPDM) under LOCA condition. *Nucl. Eng. Technol.* **2011**, *43*, 279–286. [CrossRef]
7. Šarac, T.; Devaux, J.; Quiévy, N.; Gusarov, A.; Konstantinović, M.J. The correlation between elongation at break and thermal decomposition of aged EPDM cable polymer. *Radiat. Phys. Chem.* **2017**, *132*, 8–12. [CrossRef]
8. Salehi, S.; Ezeakacha, C.P.; Kwatia, G.; Ahmed, R.; Teodoriu, C. Performance verification of elastomer materials in corrosive gas and liquid conditions. *Polym. Test.* **2019**, *75*, 48–63. [CrossRef]
9. Wang, Z.-N.; Shen, S.-L.; Zhou, A.-N.; Xu, Y.-S. Experimental evaluation of aging characteristics of EPDM as a sealant for undersea shield tunnels. *J. Mater. Civ. Eng.* **2020**, *32*, 04020182. [CrossRef]
10. Mitra, S.; Ghanbari-Siahkali, A.; Kingshott, P.; Rehmeier, H.K.; Abildgaard, H.; Almdal, K. Chemical degradation of crosslinked ethylene-propylene-diene rubber in an acidic environment. Part I. Effect on accelerated sulphur crosslinks. *Polym. Degrad. Stab.* **2006**, *91*, 69–80. [CrossRef]
11. Awwad, A.; McDaniel, D.; Lagos, L.; Rivera, J.; Tansel, B. Effect of temperature and aging duration on ethylene propylene diene monomer (EPDM) nonmetallic components used in caustic liquid waste transfer lines. *Eng. Fail. Anal.* **2021**, *181*, 105633. [CrossRef]
12. Younan, A.F.; Choneim, A.M.; Tawfik, A.A.A.; Abd-El-Nour, K.N. Electrical and physical properties of ethylene—propylene—diene monomer (EPDM) rubber loaded with semi-reinforcing furnace black. *Polym. Degrad. Stab.* **1995**, *49*, 215–222. [CrossRef]
13. Delor-Jestin, F.; Lacoste, J.; Barrois-Oudin, N.; Cardinet, C.; Lemaire, J. Photo-, Thermal and natural ageing of ethylene-propylene-diene monomer (EPDM) rubber used in automotive applications. Influence of carbon black, crosslinking and stabilizing agents. *Polym. Degrad. Stab.* **2000**, *67*, 469–477. [CrossRef]
14. Vijayalekshmi, V.; Abdul Majeed, S.S.M. Mechanical, Thermal and electrical properties of EPDM / silicone blend nanocomposites. *Int. J. Eng. Res. Appl.* **2013**, *3*, 1177–1180.
15. Zhao, Q.; Li, X.; Gao, J.; Jia, Z. Degradation evaluation of ethylene–propylene–diene monomer (EPDM) rubber in artificial weathering environment by principal component analysis. *Mater. Lett.* **2009**, *63*, 116–117. [CrossRef]
16. Li, C.; Ding, Y.; Yang, Z.; Yuan, Z.; Ye, L. Compressive stress-thermo oxidative ageing behaviour and mechanism of EPDM rubber gaskets for sealing resilience assessment. *Polym. Test.* **2020**, *84*, 106366. [CrossRef]
17. Sirisinha, K.; Chimdist, S. Comparison of techniques for determining crosslinking in silane-water crosslinked materials. *Polym. Test.* **2006**, *25*, 518–526. [CrossRef]
18. de Melo, R.P.; de Oliveira Aguiar, V.; de Fátima Vieira Marques, M. Silane crosslinked polyethylene from different commercial PE's: Influence of comonomer, catalyst type and evaluation of HLPB as crosslinking coagent. *Mater. Res.* **2015**, *18*, 313–319. [CrossRef]
19. Bansod, N.D.; Kapgate, B.P.; Maji, P.K.; Bandyopadhyay, A.; Das, C. Functionalization of Epdm rubber toward better silica dispersion and reinforcement. *Rubber Chem. Technol.* **2019**, *92*, 219–236. [CrossRef]
20. Mitra, S.; Ghanbari-Siahkali, A.; Kingshott, P.; Rehmeier, H.K.; Abildgaard, H.; Almdal, K. Chemical degradation of crosslinked ethylene-propylene-diene rubber in an acidic environment. Part II. Effect of peroxide crosslinking in the presence of a coagent. *Polym. Degrad. Stab.* **2006**, *91*, 81–93. [CrossRef]
21. Ilic, B.; Mitrovic, A.; Milicic, L. Thermal treatment of kaolin clay to obtain metakaolin. *Hem. Ind.* **2010**, *64*, 351–356. [CrossRef]
22. Blackmore, P.D.; Birtwhistle, D.; Cash, G.A.; George, G.A. Condition assessment of EPDM composite insulators using FTIR spectroscopy. *IEEE Trans. Dielectr. Electr. Insul.* **1998**, *5*, 132–141. [CrossRef]
23. Kissinger, H.E. Variation of peak temperature with heating rate in differential thermal analysis. *J. Res. Natl. Bur. Stand.* **1956**, *57*, 217. [CrossRef]
24. Sánchez-Jiménez, P.E.; Criado, J.M.; Pérez-Maqueda, L.A. Kissinger kinetic analysis of data obtained under different heating schedules. *J. Therm. Anal. Calorim.* **2008**, *94*, 427–432. [CrossRef]

25. Wang, R.; He, Y.; Wang, Z.; Kang, H.; Ma, J.; Jin, H. Effect of thermal aging on breakdown strength of EPDM rubber. *IOP Conf. Series: Mater. Sci. Eng.* **2020**, *768*, 062091. [CrossRef]
26. Woo, C.S.; Choi, S.S.; Lee, S.B.; Kim, H.S. Useful lifetime prediction of rubber components using accelerated testing. *IEEE Trans. Reliab.* **2010**, *59*, 11–17. [CrossRef]
27. Kömmling, A.; Jaunich, M.; Wolff, D. Revealing effects of chain scission during ageing of EPDM rubber using relaxation and recovery experiment. *Polym. Test.* **2016**, *56*, 261–268. [CrossRef]
28. Kömmling, A.; Jaunich, M.; Goral, M.; Wolff, D. Insights for lifetime predictions of O-ring seals from five-year long-term aging tests. *Polym. Degrad. Stab.* **2020**, *179*, 109278. [CrossRef]
29. Chailan, J.-F.; Boiteux, G.; Chauchard, J.; Pinel, B.; Seytre, G. Viscoelastic and dielectric study of thermally aged ethylene—propylene diene monomer (EPDM) compounds. *Polym. Degrad. Stab.* **1995**, *47*, 397–403. [CrossRef]
30. Zielińska, M.; Seyger, R.; Dierkes, W.K.; Bieliński, D.; Noordermeer, J.W.M. Swelling of EPDM rubbers for oil-well applications as influenced by medium composition and temperature II.experimental part. *Elastomery* **2016**, *20*, 9–24. Available online: http://yadda.icm.edu.pl/yadda/element/bwmeta1.element.baztech-6f8a523f-c7f0-4867-ae96-7811c26b12ad (accessed on 20 September 2021).
31. Srisang, W.; Supap, T.; Idem, R.; Tontiwachwuthikul, P. Study of physical and chemical resistance of elastomers in aqueous MEA and MEA+$CO_2$ solutions during the carbon dioxide absorption process. *Energy Procedia* **2014**, *63*, 1415–1423. [CrossRef]
32. Patel, H.; Salehi, S.; Ahmed, R.; Teodoriu, C. Review of elastomer seal assemblies in oil & gas wells: Performance evaluation, failure mechanisms, and gaps in industry standards. *J. Pet. Sci. Eng.* **2019**, *179*, 1046–1062. [CrossRef]

materials

Article

# A Multiphysics Peridynamic Model for Simulation of Fracture in Si Thin Films during Lithiation/Delithiation Cycles

**Xiaofei Wang and Qi Tong ***

Department of Aeronautics and Astronautics, Fudan University, Shanghai 200433, China; 19110290008@fudan.edu.cn
* Correspondence: tongqi@fudan.edu.cn

**Abstract:** Material failure is the main obstacle in fulfilling the potential of electrodes in lithium batteries. To date, different failure phenomena observed experimentally in various structures have become challenging to model in numerical simulations. Moreover, their mechanisms are not well understood. To fill the gap, here we develop a coupled chemo-mechanical model based on peridynamics, a particle method that is suitable for simulating spontaneous crack growth, to solve the fracture problems in silicon thin films due to lithiation/delithiation. The model solves mechanical and lithium diffusion problems, respectively, and uses a coupling technique to deal with the interaction between them. The numerical examples of different types of Si films show the advantage of the model in this category and well reproduce the fracture patterns observed in the experiments, demonstrating that it is a promising tool in simulating material failure in electrodes.

**Keywords:** lithium-ion batteries; electrodes; silicon thin films; fracture; peridynamics

**Citation:** Wang, X.; Tong, Q. A Multiphysics Peridynamic Model for Simulation of Fracture in Si Thin Films during Lithiation/Delithiation Cycles. *Materials* **2021**, *14*, 6081. https://doi.org/10.3390/ma14206081

Academic Editor: Dong-Joo Kim

Received: 21 August 2021
Accepted: 12 October 2021
Published: 14 October 2021



## 1. Introduction

Silicon, owing to its high energy capacity, is widely used in lithium-ion batteries as one of the most promising electrode materials. The theoretical capacity can be as high as 4200 mAh/g [1,2]. However, the current technology is still far from reaching this potential in practice. The main challenge resides in the question of how to preserve the capacity after a large number of charging/discharging cycles. The insertion and extraction of lithium can change the volume of silicon by 300% [3], leading to fragmentation of the electrodes and substantial capacity loss [4,5]. Although various nano-structured electrodes, such as nanoparticles [6,7], nanowires [2], and core-shell nanocomposites [8–10], have been fabricated to accommodate large deformation and to provide fracture resistance, mechanical degradation is still the main barrier to fulfilling the potential of silicon electrodes. Understanding the fundamental mechanism of the interplay between the lithium diffusion and the mechanical behaviors of these electrodes is desirable in guiding the design of next-generation electrodes.

The past decades have witnessed enormous efforts in studying lithium-induced material failure in silicon electrodes, experimentally and theoretically. Early works, dating back to the late 1990s, identified large volume changes as a major issue in preventing the development of high-capacity electrodes [11,12]. In situ experimental observations have found some evidence of stress evolution and fracture pattern formation due to the lithiation/delithiation process in nanostructured electrodes [13–17]. Theoretical attempts to link the mechanical stress and lithium diffusion were started by Christensen and Newman [18]. Subsequently, extensive models were built to deal with different aspects of the problem, such as the treatment of elastic-plastic material behavior [19–21] and the introduction of dislocation [22]. In addition, analytical solutions for various composite nanostructures have been developed, such as core-shell structures [23–28]. Despite this progress, in situ experiments are largely restricted in monitoring the dynamic process of chemical reactions and mechanical responses, yet they are expensive and time-consuming. On the other hand, theoretical modeling has met with difficulty in describing material failure, such as the pulverization of the electrodes.

The advancement of numerical methods provides powerful, portable and inexpensive tools for describing and predicting the chemo-mechanical interplay in silicon electrodes. Among them, cohesive zone models integrated into the finite element framework are widely used to consider the coupled problems of lithium diffusion and mechanical deformation [29,30]. Phase field models have also found success in simulating the damage evolution as lithium ions are inserted [31,32]. However, capturing the spontaneous crack growth and revealing the fracture patterns during lithiation/delithiatin cycles are still challenging for the current models. For instance, the simulation of dry bed-lake fractures in silicon thin films [33,34] requires sophisticated tools. The recently developed peridynamics method [35–37] provides a promising approach in dealing with complex fracture problems due to chemo-mechanical coupling. The model requires no supplementary criterion for crack growth, which is more convenient in this category compared with other numerical methods. The model has been successfully used in thermal and diffusion-induced deformation and fracture [38–40], multiscale crack growth [41,42], etc.

In this work, we aim at establishing a concurrently coupled chemo-mechanical model based on peridynamics to solve fracture problems in lithiated silicon thin films. Section 2 provides the basic theory of the model, including the theory of mechanical behaviors and lithium diffusion, and describes the technique that couples the two fields. Section 3 presents the numerical examples of fractures in silicon thin films due to lithiation/delithiation. Section 4 summarizes the work.

## 2. Method and Model

The peridynamic (PD) theory, introduced as a reformulation of classical continuum mechanics, is developed using integral equations, allowing the expression of damage initiation and propagation at multiple sites with arbitrary paths inside the materials without resorting to special crack growth criteria. In the PD theory, the space area is discretized into a series of material points, with each material point $x$ within a certain distance $\delta$. For material points outside this domain, it is assumed that the interactions are weak so that they can be ignored. Hence, all material points inside the domain build up the horizon ($H_x = \{x' \in R : |x' - x| \leq \delta\}$) of the material point $x$. The lines between material points $x$ and $x'$ represent PD bonds, transferring the pair-wise information between points. In peridynamic mechanical models, bonds exert forces between points in the horizon, and are used in the formulation for material deformation and damage. In the peridynamic diffusion model, bonds carry concentration information from one point to another.

In this section, we briefly describe the peridynamic mechanical and diffusion models. We then introduce the newly developed chemo-mechanical model to study spontaneous crack growth in silicon thin films during lithiation/delithiation processes.

### 2.1. Bond-Based Peridynamic Mechanical Model

The PD equation of motion of each material point can be written according to Newton's Second Law as

$$\rho \frac{\partial^2 u(x,t)}{\partial t^2} = \int_{H_x} f(u(x',t), u(x,t), x', x, t) dV_{x'} + b(x,t) \tag{1}$$

where $\rho$ is the mass density, $u$ is the displacement, $dV_{x'}$ is the integration variable that indicates the infinitesimal domain located at point $x'$, and $H_x$ is the horizon, representing the family of particles within a local region centered by $x$. $b$ is a prescribed body-force-density field, $f$ is the response function which is defined as the force vector per unit volume squared that the material point $x'$ exerts on the material point at $x$ if their distance is less than the radius $\delta$ of the horizon. The undeformed bond vector is defined as $\varepsilon = x' - x$, and the relative displacement is $\eta = u' - u$. Thus, the deformed bond vector can be calculated as $\eta + \varepsilon = (x' + u') - (x + u)$.

In the PD theory, the interaction between two material points is described by the force density function. All material-specific information is contained in the force density function that illustrates the dependence between the interparticle forces on the reference positions and the displacements of the points. The displacement of a material point is then integrated based on Equation (1). The force density function for a linear micro-elastic material is given as

$$f(\eta, \varepsilon) = \frac{\eta + \varepsilon}{|\eta + \varepsilon|} u(\varepsilon, t) c s \tag{2}$$

where $c$ is the micro-modulus function of the bond. For three-dimensional problems, the micro-modulus function of the bond can be defined as

$$c = \frac{12E}{\pi \delta^4} \tag{3}$$

Here $E$ is Young's modulus, $\delta$ is the size of the horizon. $s$ is the stretch that is defined as

$$s = \frac{|\eta + \varepsilon| - |\varepsilon|}{|\varepsilon|} \tag{4}$$

and it is the ratio of the change in distance to the initial distance between points $x'$ and $x$ . The failure condition of each bond can be represented by a failure parameter as

$$\mu(\varepsilon, t) = \begin{cases} 1 & s < s_0 \\ 0 & s \geq s_0 \end{cases} \tag{5}$$

where $s_0$ is the critical stretch, which can be calculated in terms of the critical energy release rate of a particular material. For three-dimensional problems, the critical stretch of the bond can be defined as

$$s_0 = \sqrt{\frac{5\Gamma}{9K\delta}} \tag{6}$$

where $\Gamma$ is the fracture energy and $K$ is the bulk modulus. When the stretch of the bond between two points exceeds the critical stretch $s_0$, the bond breaks, and these two points cease to interact with each other. Once the bond breaks, the load on material point $x$ will be redistributed to the remaining bonds and the deformation state changes. Therefore, damage is treated as part of the constitutive model through the irreversible breakage of interactions. In order to quantify the damage state of a material point, a local damage parameter of a material point is defined as

$$\varphi(x, t) = 1 - \frac{\int_{H(x)} \mu(\varepsilon, t) dV}{\int_{H(x)} dV} \tag{7}$$

Local damage is the percentage of broken bonds associated with the material point $x$ and varies from 0 to 1, where 0 represents undamaged and 1 refers to fully damaged cases.

### 2.2. *Peridynamic Diffusion Model*

In order to model lithium diffusion, we introduce the diffusion equation that characterize the mass transport

$$\frac{\partial C}{\partial t} + \nabla \cdot J = 0 \tag{8}$$

where the Li ion concentration $C$ is the number of Li atoms per unit volume. The boundary value is represented by $C_0$. The flux of the lithium ions $J$ is the number of Li atoms per unit time crossing a unit area, which consists of two contributions in a chemo-mechanical coupled environment [43]. The first term is due to the concentration gradient,

$$J_{Li,c} = -D_0 \nabla C \tag{9}$$

The other contribution is from the stress,

$$J_{Li,\sigma} = D_0(1 - \frac{C}{C_{max}})\frac{\Omega_{Li}C}{RT}\nabla\sigma \tag{10}$$

The hydrostatic stress gradient that drives Li mass transport reaches zero at both $C_0 = 0$ and $C_0 = C_{max}$. In the above equations, $D_0$ is the diffusivity of lithium ions, $\Omega$ is the partial molar volume of Li in silicon, $k$ is the gas constant, $T$ is the temperature, and $\sigma_m = (\sigma_1 + \sigma_2 + \sigma_3)/3$ is the bulk stress, with $\sigma_1$, $\sigma_2$, and $\sigma_3$ being diagonal terms of the Cauchy stress tensor. The Cauchy stress tensor is calculated as

$$\sigma = \frac{1}{2}\int_{H(x)} f(\eta,\varepsilon) \otimes (\eta + \varepsilon)dV_{x'} \tag{11}$$

The dynamic process of lithium diffusion is represented by Fick's second law. Note that the stress induced by a large volume change during cycling cannot be ignored. By considering the concentration gradient and hydrostatic stress gradient, Fick's second law is modified as

$$\frac{\partial C(x,t)}{\partial t} = D_0\frac{\partial^2 C}{\partial r^2} - \frac{D_0\Omega}{kT}\left[C\left(1 - \frac{C}{C_{max}}\right)\frac{\partial^2\sigma_m}{\partial r^2} + \left(1 - \frac{2C}{C_{max}}\right)\frac{\partial\sigma_m}{\partial r} \times \frac{\partial C}{\partial r}\right] \tag{12}$$

Although peridynamics was originally introduced for deformation fields, it is also applicable to other fields, including diffusion. However, it is not straightforward to express the coupled diffusion equation in a peridynamic form. In this study, this has been achieved using the concept of the peridynamic differential operator [44–46]. All spatial derivatives in the classical diffusion equation can be transformed into peridynamic form as

$$\frac{\partial^2 C(x,t)}{\partial r^2} = \left(\frac{6}{\pi\delta^4}\right)\int_{H(x)}\frac{C(x',t) - C(x,t)}{|\varepsilon|}dV_{x'} \tag{13}$$

$$\frac{\partial^2 \sigma_m(x,t)}{\partial r^2} = \left(\frac{6}{\pi\delta^4}\right)\int_{H(x)}\frac{\sigma_m(x',t) - \sigma_m(x,t)}{|\varepsilon|}dV_{x'} \tag{14}$$

$$\begin{aligned}\frac{\partial \sigma_m(x,t)}{\partial r} \times \frac{\partial C(x,t)}{\partial r} =& \left(\frac{81}{16\pi^2\delta^6}\right)\left(\int_{H(x)}\frac{\sigma_m(x',t) - \sigma_m(x,t)}{|\varepsilon|}dV_{x'}\right) \\ &\times \left(\int_{H(x)}\frac{C(x',t) - C(x,t)}{|\varepsilon|}dV_{x'}\right)\end{aligned} \tag{15}$$

Therefore, the general Fick's second law given in Equation (12) can be written in peridynamic form as

$$\begin{aligned}\frac{\partial C(x,t)}{\partial t} =& D_0\frac{\partial^2 C}{\partial r^2} - \frac{D_0\Omega}{kT}\left[C\left(1 - \frac{C}{C_{max}}\right)\frac{\partial^2\sigma_m}{\partial r^2} + \left(1 - \frac{2C}{C_{max}}\right)\frac{\partial\sigma_m}{\partial r} \times \frac{\partial C}{\partial r}\right] \\ =& \left(\frac{6D_0}{\pi\delta^4}\right)\int_{H(x)}\frac{C(x',t) - C(x,t)}{|x' - x|}dV_{x'} \\ &- \left(\frac{6D_0\Omega}{\pi\delta^4 kT}\right)\int_{H(x)}C(x,t)\left(1 - \frac{C(x,t)}{C_{max}}\right)\frac{\sigma_m(x',t) - \sigma_m(x,t)}{|x' - x|}dV_{x'} \\ &- \left(\frac{81D_0\Omega}{16\pi^2\delta^6 kT}\right)\left(\int_{H(x)}\frac{\sigma_m(x',t) - \sigma_m(x,t)}{|x' - x|}dV_{x'}\right) \\ &\times \left(\int_{H(x)}\left(1 - \frac{2C(x,t)}{C_{max}}\right)\frac{C(x',t) - C(x,t)}{|x' - x|}dV_{x'}\right)\end{aligned} \tag{16}$$

### 2.3. Coupled Peridynamic Chemo-Mechanical Model

This study also extends the peridynamic theory to include the effect of chemo-mechanical loading. Attributed to the insertion of lithium ions, the silicon electrodes experience huge volume changes during the processes of lithiation and delithiation. Therefore, it is necessary to modify the force density function and the failure parameter, i.e., to subtract the volume expansion caused by the intercalation of lithium ions. Without considering the influence of Li-ion concentration, the stretch of the bond between the two material points is $s = (|\eta + \varepsilon| - |\varepsilon|)/|\varepsilon|$. When the influence of the Li ion concentration is considered, lithiation-induced linear strain is

$$s_{Li} = \frac{1}{3}\log\left(1 + \frac{\Omega(C(x,t) + C(x',t))}{2}\right) \tag{17}$$

Furthermore, a new response function is proposed to include the effects of chemical loading and the distance between the points as

$$\begin{aligned} f(\eta,\varepsilon) &= \frac{\eta+\varepsilon}{|\eta+\varepsilon|}\mu(\varepsilon,t)c(s - s_{Li}) \\ &= \frac{\eta+\varepsilon}{|\eta+\varepsilon|}\mu(\varepsilon,t)c\left(\frac{|\eta+\varepsilon| - |\varepsilon|}{|\varepsilon|} - \frac{1}{3}\log\left(1 + \frac{\Omega(C(x,t) + C(x',t))}{2}\right)\right) \end{aligned} \tag{18}$$

Similarly, the effect of chemical loading is also included in the failure parameter by extending its definition, given in Equation (5), as

$$\mu(\varepsilon,t) = \begin{cases} 1 & \frac{|\eta+\varepsilon|-|\varepsilon|}{|\varepsilon|} - \frac{1}{3}\log\left(1 + \frac{\Omega\left(C(x,t)+C(x',t)\right)}{2}\right) < s_0 \\ 0 & \frac{|\eta+\varepsilon|-|\varepsilon|}{|\varepsilon|} - \frac{1}{3}\log\left(1 + \frac{\Omega\left(C(x,t)+C(x',t)\right)}{2}\right) \geq s_0 \end{cases} \tag{19}$$

in which $s_0$ is still the critical stretch upon failure. However, the failure between two points occurs when the elastic stretch $s_0 - s_{Li}$, rather than stretch $s$, exceeds the critical stretch $s_0$.

### 2.4. Flowchart of the Numerical Scheme

To simulate the spontaneous crack growth in Si thin films, we developed the algorithm shown in Figure 1. The coupled chemo-mechanical model features a large time-scale separation between the diffusion and the fracture processes. Typically, the time scale of the diffusion process is several orders higher than that of the force field. To reduce the computational cost, we employ a quasi-static solver for the force field, whereas the diffusion follows a dynamic time evolution.

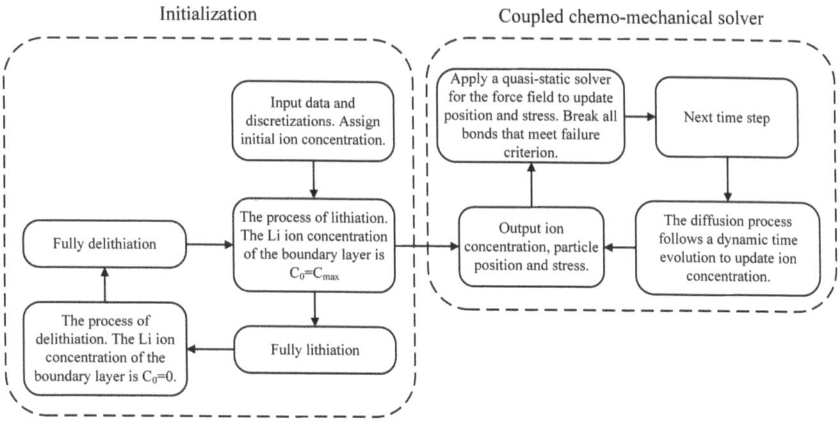


**Figure 1.** Flowchart of the numerical scheme.

## 3. Numerical Examples

### 3.1. Numerical Model

The model investigated is an Si thin film deposited on the TiC substrate, as shown in Figure 2. The Si thin film has the width $L$ and the thickness $H$. The TiC substrate has the width $L_1$ and the thickness $H_1$. The lithium ions are allowed to penetrate the Si thin film over the entire surface that is in contact with the electrolyte. A prescribed constant concentration of lithium ions is considered over the upper free surface. The free surface in contact with the electrolyte has a stress-free condition applied, allowing for volume expansion, whereas the other surface is deposited on the TiC substrate, restricted by the interface. In its present form, the resolution of the diffusion of the lithium ions into the Si thin film is not coupled with its mechanical state, meaning that stress does not assist the ions' diffusion. The boundary conditions of the Si thin film are illustrated in Figure 2. The diffusion coefficient of lithium ions in the silicon is $2.5 \times 10^{-17}$ m$^2$/s. The Young's moduli of the lithiated silicon and the TiC substrate are $E_{Si} = 12$ GPa and $E_{TiC} = 439.4$ GPa, respectively. An effective modulus is introduced to account for the interface, given by $1/E_e = (1/E_{Si} + 1/E_{TiC})/2$. The fracture energy is $\Gamma_f = 2000$ J/m$^2$ for silicon in the first lithiation. During the delithiation process, which consists of broken atomic bonds, the fracture energy was taken to be five times smaller than that during lithiation, i.e., $\Gamma_f = 400$ J/m$^2$ for silicon in the first delithiation. Furthermore, $\Gamma_d = 500$ J/m$^2$ is used for the $Si - TiC$ interface. To model the progressive degradation of the material properties after several cycles, the fracture energy $\Gamma_f$ is decreased by 9% of its initial value after each cycle of lithiation and delithiation. In the tenth cycle, the fracture energy has been decreased by about 80% of its initial value. The complete list of the parameters of the materials and the numerical model is provided in Table 1.

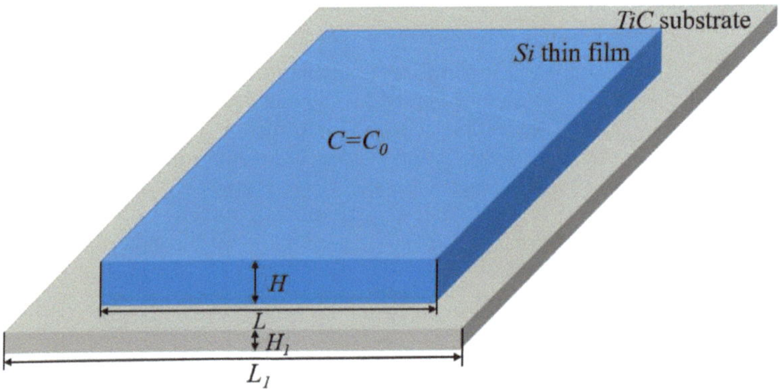


**Figure 2.** The Si thin film was deposited on the TiC substrate and the boundary conditions for the analysis of the Si thin film in the processes of lithiation and delithiation.

Experimental observations of the fracture patterns in different silicon membranes are shown in Figure 3. The most common fracture pattern is dry lake-bed cracks [32], which is illustrated in Figure 3a. This type of fracture is typically observed in continuous flat membranes. Since the in-plane stress is isotropic during volume expansion when lithium ions are inserted, cracks grow without a favorable direction. As a result, the crack paths are fairly random. Other types of membranes, shown in Figure 3b,c, are patterned with holes to accommodate volume expansion and to mitigate internal stress. The fracture patterns are more regular and usually crosslink the holes. The proposed multiphysics model was validated in these membranes. The experimental techniques were found in the literature, e.g., [32]. We followed similar procedures in the simulations.

**Table 1.** Parameters of the materials and the numerical model.

| Parameter | Symbol | Value | Units |
|---|---|---|---|
| Diffusion coefficient | $D_0$ | $2.5 \times 10^{-17}$ | $m^2/s$ |
| Young's modulus of lithiated Si | $E_{Si}$ | 12 | GPa |
| Young's modulus of TiC | $E_{TiC}$ | 439.4 | GPa |
| Young's modulus of Si-TiC | $E_e = \frac{2}{(\frac{1}{E_{Si}} + \frac{1}{E_{TiC}})}$ | 23.36 | GPa |
| Poisson's ratio | $\upsilon$ | 0.25 | |
| Fracture energy of Si in first lithiation | $\Gamma_f$ | 2000 | $J/m^2$ |
| Fracture energy of Si in first delithiation | $\Gamma_f$ | 400 | $J/m^2$ |
| Fracture energy of Si-TiC | $\Gamma_d$ | 500 | $J/m^2$ |
| Partial molar volume of Si | $\Omega$ | $1.2052 \times 10^{-5}$ | $m^3/mol$ |
| Gas constant | $k$ | 8.314 | |
| Temperature | $T$ | 300 | K |
| Width of Si thin film | $L$ | $1.6 \times 10^4$ | nm |
| Thickness of Si thin film | $H$ | 600 | nm |
| Width of TiC substrate | $L_1$ | $2.0 \times 10^4$ | nm |
| Thickness of TiC substrate | $H_1$ | 300 | nm |
| Grid size | $\Delta$ | 100 | nm |
| Horizon | $\delta$ | 300 | nm |
| Critical stretch of Si in first lithiation | $s_0 = \sqrt{\frac{5\Gamma_f}{9K_{Si}\delta}}$ | 0.68 | |
| Critical stretch of Si in first delithiation | $s_0 = \sqrt{\frac{5\Gamma_f}{9K_{Si}\delta}}$ | 0.304 | |
| Critical stretch of Si-TiC | $s_0 = \sqrt{\frac{5\Gamma_d}{9K_e\delta}}$ | 0.244 | |

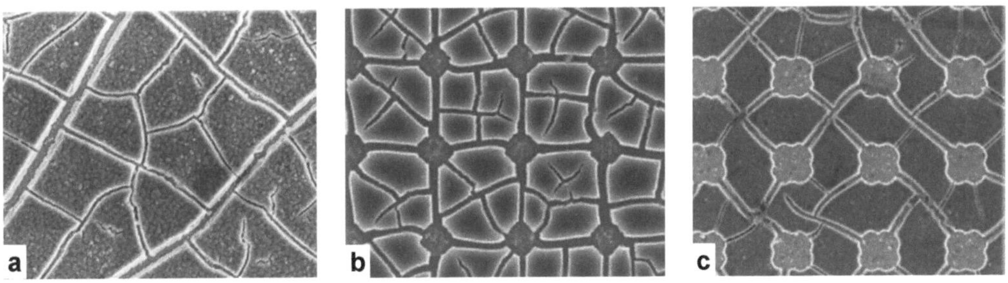


**Figure 3.** Fracture patterns in (**a**) continuous silicon thin film, (**b**,**c**) patterned films. Reproduced with permission [32].

### 3.2. Fracture of Continuous Membrane

Figure 4 shows snapshots of crack formation and propagation in the continuous silicon membrane during lithiation/delithiation cycles. To investigate the effect of the charging/discharging rate, we considered the cases including one and two cycles in 60,000 s, respectively. Note that the fracture energy $\Gamma_f$ decreased by 9% in each cycle because of the material degradation. Figure 4a presents the case with one cycle in 60,000 s. The snapshots are viewed from the interface instead of the free surface to better observe the cracks. The particles are colored based on the damage parameter in Equation (7). The process of lithiation was in the period of 0–30,000 s. Crack initiation is observed at $t$ = 15,000 s at some random locations. The volume expansion led to partial detachment of the film from the substrate, through breaking the bonds between the particles in the film and the substrate. As a result, the contour of the film became irregular due to the combined effect of detachment and volume expansion. By the end of the lithiation at $t$ = 30,000 s, apparent crack lines were formed, and debonding was propagated inward from each side observed from the damage profile. The inset at $t$ = 30,000 s shows the free surface at the same time. It seems that the cracks did not propagate through the thickness direction of the film during lithiation. This was the consequence of the bending upward of the film, where the interface was mostly subjected to tensile stress and the free surface was compressed, where cracks were suppressed. The process of delithiation was in the period of 30,000–60,000 s. The crack segments formed during lithiation further propagated to connect with each other. A crack network had formed at $t$ = 45,000 s on the interface and started to travel through the thickness direction of the film. By the end of the cycle at $t$ = 60,000 s, apparent fragmentation through the thickness direction of the film could be observed. The overall fracture pattern was isotropic without a preferable crack direction, which corresponds to the dry lake-bed cracks in Figure 3a.

Figure 4b further illustrates the case of two cycles in 60,000 s. Each lithiation or delithiation took 15,000 s. Apparent differences were found as the charging/discharging rate was doubled. By the end of the first cycle at $t$ = 30,000 s, crack segments formed but did not crosslink. This feature can be attributed to the faster charging/discharging rate that released and redistributed the internal stress more frequently. Redistribution of stress provided a tendency to form more bifurcations instead of directly linking to each other. In contrast, in the previous case, the lithiation process took 30,000 s, which was enough for the crack segments to become connected under high stress intensity. On the other hand, observable in the insets (free surface) at $t$ = 30,000 s, the cracks had already traveled through the thickness direction of the film. This can be compared with the snapshot at the same time in the previous case, where the free surface was still undamaged. Therefore, a complete cycle of lithiation/delithiation was desirable to provide enough tensile stress for the propagating of the cracks through the thickness direction. This was evidenced at $t$ = 60,000 s in the previous case. If the cycles kept running, more and more bifurcations would be generated due to the degradation of the materials.

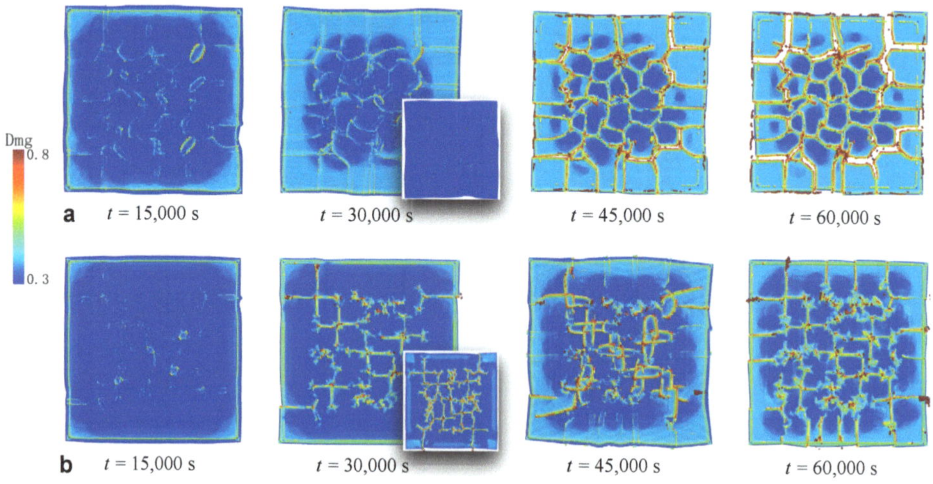


**Figure 4.** The processes of crack formation and propagation in the continuous Si film during lithiation/delithiation cycles. (**a**) Snapshots in the case of one cycle in 60,000 s. (**b**) Snapshots in the case of two cycles in 60,000 s. The snapshots depict the view from the interface on the bottom of the film. The insets at $t$ = 30,000 s are snapshots on the top free surface at the same moments. Colors represent the degrees of damage.

### 3.3. Fracture of Membrane with Square Holes

To study the fracture in the patterned Si thin films in Figure 3b,c, we created a new sample with nine square holes evenly distributed at 25 μm, 75 μm, and 125 μm in each direction from the corners, as shown in Figure 5a. The width of the squares was 25 μm. The snapshots were taken from the interface, as in the continuous film. The damage profiles on the top free surface are also presented in the insets at the end of the lithiation/delithiation in Figure 5c,f. The coloring method is the same as before.

Several implications can be drawn from the period of lithiation shown in Figure 5b,c. First, the holes were greatly contracted compared with the original configuration. This is the advantage of this structure, where the holes can accommodate the strain caused by lithium-ion insertion through the expansion inward. Therefore, internal stress was mitigated. Second, cracks and debonding formed more easily at the corners of the film as well as the holes due to stress concentration. The initial cracks tended to connect to each other, resulting in diagonally crosslinked crack paths. Third, in contrast to the continuous film, the patterned film demonstrated a regular fracture pattern and a deformed shape. As was the case with the continuous film, cracks were initiated on the bottom interface and did not travel through the film in the thickness direction during lithiation, which is evidenced in the inset in Figure 5c. The followed delithiation process is illustrated in Figure 5d,f. As the lithium ions were extracted from the film, the shape of the film recovered, and the holes turned back to their original sizes. However, the damage was intensified as the cracks walked through the film, as demonstrated in the inset in Figure 5f. The diagonalized fracture pattern can be compared with the experimental results in Figure 3c.

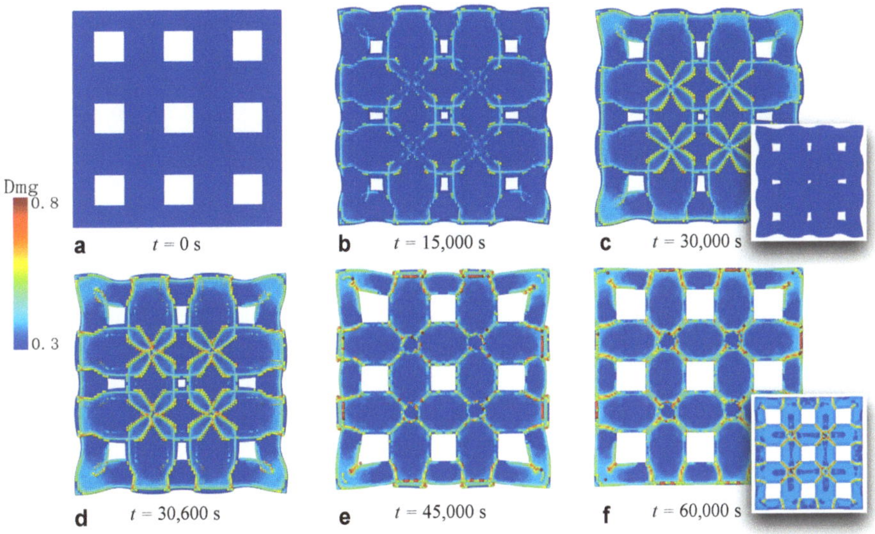


**Figure 5.** The process of crack formation and propagation in the silicon film pattered with square holes. The holes are located at 25 μm, 75 μm, and 125 μm in each direction from the corners. The width of the holes is 25 μm. The figures and the insets correspond to the same process of lithiation/delithiation as in the continuous film.

### *3.4. Fracture of Membrane with Circular Holes*

The third numerical simulation was conducted based on a silicon thin film patterned with circular holes. The locations and the sizes of the holes were the same as the square holes in the previous sample. The lithiation/delithiation processes are illustrated in Figure 6, where a–c depict the lithiation from 0 to 30,000 s, and d–f present the period of delithiation. Similarly to the continuous and square-patterned films, the cracks initiated and propagated from the interface during lithiation and did not run through the thickness direction. The holes also experienced significant shrinkage as holes of irregular shape and different sizes. However, the cracks grew along the horizontal and vertical directions, and then developed a network by connecting to each other. Since the circular holes do not have sharp corners to induce stress concentration, diagonal connections are not favorable compared with the shorter lines in horizontal and vertical directions. It is noteworthy that the hole in the middle produced cracks along the diagonal directions, but these cracks did not fully expand and connect. Compared with the film with square holes, circular holes provide better accommodation for volume expansion. Throughout the lithiation/delithiation cycle, the overall shape was largely conserved, without significant distortion.

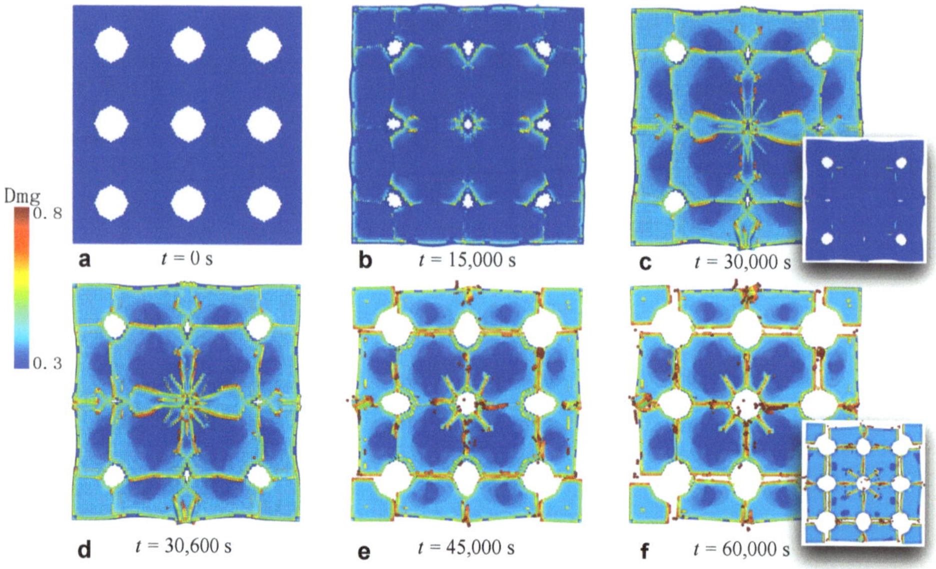


**Figure 6.** The process of crack formation and propagation in the silicon film pattered with circular holes. The holes were located at 25 μm, 75 μm, and 125 μm in each direction from the corners. The diameter of the holes was 25 μm. The figures and the insets correspond to the same process of lithiation/delithiation as in the continuous film.

## 4. Conclusions

In summary, we have developed a concurrently coupled chemo-mechanical model based on peridynamics to simulate the complex fracture problems in different types of silicon thin films due to the insertion and extraction of lithium ions. The framework consists of a classic bond-based peridynamic mechanical model, a diffusion model and a coupling technique to reflect the interaction between them. We have used the method to perform numerical simulations of three different thin films, i.e., continuous film and films patterned with square and circular holes, which are typically used to mitigate volume expansion and to reduce the risk of material failure. The continuous film displayed a dry lake-bed type fracture pattern, i.e., an isotropic crack network without a favorable direction. We also investigated the influence of the charging/discharging rate and revealed that a higher rate resulted in more bifurcations because the internal stress was released and redistributed more frequently. The cracks in the patterned films formed networks that connected the holes. The square holes and circular holes experienced different styles of connection, which were diagonal or horizontal/vertical. The simulations reproduced the fracture patterns observed in experiments, which proves the capability of the model to deal with the problems of material failure in electrodes, and to provide a powerful tool to aid in the future design of new structures. Future work includes the applications of the model in different structures and materials, such as spherical particles, hollow core–shell nanotubes, Si electrodes that contain carbon additives and polymer binders, etc. It is also desirable to incorporate complex constitutive models into the framework to study the influence of crystallography.

**Author Contributions:** Conceptualization, writing, supervision, and project administration, Q.T.; investigation, coding, and data processing, X.W. All authors have read and agreed to the published version of the manuscript.

**Funding:** This research was funded by NSFC grant number 11902079, Shanghai Association for Science and Technology grant number 19PJ1401200 and Fudan University.

**Institutional Review Board Statement:** Not applicable.

**Informed Consent Statement:** Not applicable.

**Data Availability Statement:** The related data are available upon request.

**Conflicts of Interest:** The authors declare no conflict of interest.

## References

1. Boukamp, B.A.; Lesh, G.C.; Huggins, R.A. All-Solid Lithium Electrodes with Mixed-Conductor Matrix. *J. Electrochem. Soc.* **1981**, *128*, 725–729. [CrossRef]
2. Chan, C.K.; Peng, H.; Liu, G.; McIlwrath, K.; Zhang, X.; Huggins, R.A.; Cui, Y. High-performance lithium battery anodes using silicon nanowires. *Nat. Nanotech.* **2008**, *3*, 31–35. [CrossRef] [PubMed]
3. Zhang, W.J. A review of the electrochemical performance of alloy anodes for lithium-ion batteries. *J. Power Sources* **2011**, *196*, 13–24. [CrossRef]
4. Arora, P.; White, R.E.; Doyle, M. Capacity fade mechanisms and side reactions in lithium-ion batteries. *J. Electrochem. Soc.* **1998**, *145*, 3647. [CrossRef]
5. Wang, D.; Wu, X.; Wang, Z.; Chen, L. Cracking causing cyclic instability of LiFePO4 cathode material. *J. Power Sources* **2005**, *140*, 125–128. [CrossRef]
6. Son, D.; Kim, E.; Kim, T.G.; Kim, M.G.; Cho, J.; Park, B. Nanoparticle iron-phosphate anode material for Li-ion battery. *Appl. Phys. Lett.* **2004**, *85*, 5875. [CrossRef]
7. Park, M.H.; Kim, K.; Kim, J.; Cho, J. Flexible dimensional control of high-capacity Li-ion-battery anodes: From 0D hollow to 3D porous germanium nanoparticle assemblies. *Adv. Mater.* **2010**, *22*, 415. [CrossRef]
8. Cui, L.F.; Ruffo, R.; Chan, C.K.; Peng, H.L.; Cui, Y. Crystalline-amorphous core-shell silicon nanowires for high capacity and high current battery electrodes. *Nano Lett.* **2009**, *9*, 491–495. [CrossRef]
9. Liu, N.; Wu, H.; McDowell, M.T.; Yao, Y.; Wang, C.; Cui, Y. A yolk–shell design for stabilized and scalable Li-ion battery alloy anodes. *Nano Lett.* **2012**, *12*, 3315–3321. [CrossRef]
10. Liu, N.; Lu, Z.; Zhao, J.; McDowell, M.T.; Lee, H.W.; Zhao, W.; Cui, Y. A pomegranate-inspired nanoscale design for large-volume-change lithium battery anodes. *Nat. Nanotech.* **2014**, *9*, 187–192. [CrossRef]
11. Winter, M.; Besenhard, J.O. Electrochemical lithiation of tin and tin-based intermetallics and composites. *Electrochim. Acta* **1999**, *45*, 31–50. [CrossRef]
12. Sun, Y.; Liu, N.; Cui, Y. Promises and challenges of nanomaterials for lithium-based rechargeable batteries. *Nat Energy* **2016**, *1*, 16071. [CrossRef]
13. Liu, X.; Huang, J. In situ TEM electrochemistry of anode materials in lithium ion batteries. *Energy Environ. Sci.* **2011**, *4*, 3844–3860. [CrossRef]
14. Xie, H.; Qiu, W.; Song, H.; Tian, J. InSitumeasurementofthedeformationandelas- tic modulus evolution in Si composite electrodes during electrochemical lithiation and delithiation. *J. Electrochem. Soc.* **2016**, *163*, A2685–A2690. [CrossRef]
15. Sandu, G.; Brassart, L.; Ois Gohy, J.F.; Pardoen, T.; Melinte, S.; Vlad, A. Surface Coating Mediated Swelling and Fracture of Silicon Nanowires during Lithiation. *ACS Nano* **2014**, *8*, 9427–9436. [CrossRef]
16. Shi, F.; Song, Z.; Gabor A. Somorjai, P.N.R.; Ritchie, R.O.; Komvopoulos, K. Failure mechanisms of single-crystal silicon electrodes in lithium-ion batteries. *Nat. Commun.* **2016**, *7*, 11886. [CrossRef] [PubMed]
17. Lee, S.W.; McDowella, M.T.; Berlaa, L.A.; Nixa, W.D.; Cui, Y. Fracture of crystalline silicon nanopillars during electrochemical lithium insertion. *Proc. Natl. Acad. Sci. USA* **2012**, *109*, 4080–4085. [CrossRef]
18. Christensen, J.; Newman, J. A mathematical model of stress generation and fracture in lithium manganese oxide. *J. Electrochem. Soc.* **2006**, *153*, A1019–A1030. [CrossRef]
19. Zhao, K.; Pharr, M.; Vlassak, J.; Suo, Z. Inelastic hosts as electrodes for high-capacity lithium-ion batteries. *J. Appl. Phys.* **2011**, *109*, 016110. [CrossRef]
20. Zhao, K.; Pharr, M.; Cai, S.; Vlassak, J.; Suo, Z. Large plastic deformation in high-capacity lithium-ion batteries caused by charge and discharge. *J. Am. Ceram. Soc.* **2011**, *94*, S226–S235. [CrossRef]
21. Anand, L. A Cahn–Hilliard-type theory for species diffusion coupled with large elastic–plastic deformations. *J. Mech. Phys. Solids* **2012**, *60*, 1983–2002. [CrossRef]
22. Li, J.; Fang, Q.; Liu, F.; Liu, Y. Analytical modeling of dislocation effect on diffusion induced stress in a cylindrical lithium ion battery electrode. *J. Power Sources* **2014**, *272*, 121–127. [CrossRef]

23. Cui, Z.; Gao, F.; Qu, J. A finite deformation stress-dependent chemical potential and its applications to lithium ion batteries. *J. Mech. Phys. Solids* **2012**, *60*, 1280. [CrossRef]
24. Hao, F.; Fang, D. Diffusion-Induced Stresses of Spherical Core-Shell Electrodes in Lithium-Ion Batteries: The Effects of the Shell and Surface/Interface Stress. *J. Electrochem. Soc.* **2013**, *160*, A595–A600. [CrossRef]
25. Ma, Z.; Xie, Z.; Wang, Y.; Zhang, P.; Pan, Y.; Zhou, Y.; Lu, C. Failure modes of hollow coreeshell structural active materials during the lithiationedelithiation process. *J. Power Sources* **2015**, *290*, 114–122. [CrossRef]
26. Zhao, K.; Pharr, M.; Hartle, L.; Vlassak, J.J.; Suo, Z. Fracture and debonding in lithium-ion batteries with electrodes of hollow coreeshell nanostructures. *J. Power Sources* **2012**, *218*, 6–14. [CrossRef]
27. Xu, C.; Weng, L.; Chen, B.; Zhou, J.; Cai, R. An analytical model for the fracture behavior in hollow cylindrical anodes. *Int. J. Mech. Sci.* **2019**, *157–158*, 87–97. [CrossRef]
28. Hao, F.; Fang, D. Tailoring diffusion-induced stresses of core- shell nanotube electrodes in lithium-ion batteries. *J. Appl. Phys.* **2013**, *113*, 013507. [CrossRef]
29. Sun, G.; Sui, T.; Song, B.; Zheng, H.; Lu, L.; Korsunsky, A.M. On the fragmentation of active material secondary particles in lithium ion battery cathodes induced by charge cycling. *Extrem. Mech. Lett.* **2016**, *9*, 449–458. [CrossRef]
30. Bucci, G.; Swamy, T.; Chiang, Y.M.; Carter, W.C. Modeling of internal mechanical failure of all-solid-state batteries during electrochemical cycling, and implications for battery design. *J. Mater. Chem. A* **2017**, *5*, 19422–19430. [CrossRef]
31. Klinsmann, M.; Rosato, D.; Kamlah, M.; McMeeking, R.M. Modeling crack growth during Li insertion in storage particles using a fracture phase field approach. *J. Mech. Phys. Solids* **2016**, *92*, 313–344. [CrossRef]
32. Réthoré, J.; Zheng, H.; Li, H.; Li, J.; Aifantis, K.E. A multiphysics model that can capture crack patterns in Si thin films based T on their microstructure. *J. Power Sources* **2018**, *400*, 383–391. [CrossRef]
33. Li, J.; Dozier, A.K.; Li, Y.; Yang, F.; Cheng, Y.T. Crack Pattern Formation in Thin Film Lithium-Ion Battery Electrodes. *J. Electrochem. Soc.* **2011**, *158*, A689–A694. [CrossRef]
34. Li, T.; Gulzar, U.; Zaccaria, R.P.; Capiglia, C.; Hackney, S.A.; Aifantis, K.E. Damage Formation in Sn Film Anodes of Na-Ion Batteries. *J. Phys. Chem. C* **2019**, *123*, 15244–15250. [CrossRef]
35. Silling, S. Reformulationofelasticitytheoryfordiscontinuitiesandlong-rangeforces. *J. Mech. Phys. Solids* **2000**, *48*, 175–209. [CrossRef]
36. Silling, S.; Epton, M.; Weckner, O.; Xu, J.; Askari, E. Peridynamic States and Constitutive Modeling. *J. Elast.* **2007**, *88*, 151–184. [CrossRef]
37. Silling, S.; Lehoucq, R. Peridynamic theory of solid mechanics. *Adv. Appl. Mech.* **2010**, *44*, 73–166.
38. Chen, Z.; Bobaru, F. Peridynamic modeling of pitting corrosion damage. *J. Mech. Phys. Solids* **2015**, *78*, 352–381. [CrossRef]
39. Chen, Z.; Jafarzadeh, S.; Zhao, J.; Bobaru, F. A coupled mechano-chemical peridynamic model for pit-to-crack transition in stress-corrosion cracking. *J. Mech. Phys. Solids* **2021**, *146*, 104203. [CrossRef]
40. Shi, C.; Gong, Y.; Yang, Z.; Tong, Q. Peridynamic investigation of stress corrosion cracking in carbon steel pipes. *Eng. Fract. Mech.* **2019**, *219*, 1–11. [CrossRef]
41. Tong, Q.; Li, S. Multiscale coupling of molecular dynamics and peridynamics. *J. Mech. Phys. Solids* **2016**, *95*, 169–187. [CrossRef]
42. Tong, Q.; Li, S. A concurrent multiscale study of dynamic fracture. *Comput. Methods Appl. Mech. Eng.* **2020**, *366*, 113075. [CrossRef]
43. Bohn, E.; Eckl, T.; Kamlah, M.; McMeeking, R. A Model for Lithium Diffusion and Stress Generation in an Intercalation Storage Particle with Phase Change. *J. Electrochem. Soc.* **2013**, *160*, A1638–A1652. [CrossRef]
44. Wang, H.; Oterkus, E.; Oterkus, S. Peridynamic modelling of fracture in marine lithium-ion batteries. *Ocean Eng.* **2018**, *151*, 257–267. [CrossRef]
45. Wang, H.; Oterkus, E.; Oterkus, S. Predicting fracture evolution during lithiation process using peridynamics. *Eng. Fract. Mech.* **2018**, *192*, 176–191. [CrossRef]
46. Wang, H.; Oterkus, E.; Oterkus, S. Three-Dimensional Peridynamic Model for Predicting Fracture Evolution during the Lithiation Process. *Energies* **2018**, *11*, 1461. [CrossRef]

*materials*

MDPI

*Article*

# Influence of TIG and Laser Welding Processes of Fe-10Cr-4Al-RE Alloy Cracks Overlayed on 316L Steel Plate

Bin Xi [1], Bo Liu [1], Song Li [1], Disi Wang [1], Youpeng Zhang [1,*], Peter Szakálos [2], Jesper Ejenstam [3], Janne Wallenius [4], Guangqing He [1,5] and Wenyang Zhang [5]

1 Institute of Modern Physics, Fudan University, Shanghai 200433, China; 19110200027@fudan.edu.cn (B.X.); 20210200011@fudan.edu.cn (B.L.); 21210200010@fudan.edu.cn (S.L.); 20210200015@fudan.edu.cn (D.W.); heguangqing2013@163.com (G.H.)
2 Division of Surface Chemistry and Corrosion Science, Royal Institute of Technology, 10691 Stockholm, Sweden; peter@leadcold.com
3 Sandvik Heating Technology AB, Kanthal, 73427 Hallstahammar, Sweden; jesper.ejenstam@sandvik.com
4 Division of Nuclear Engineering, Royal Institute of Technology, 10691 Stockholm, Sweden; janwal@kth.se
5 Shanghai Electric Group, Shanghai 200233, China; zhangwy2@shanghai-electric.com
* Correspondence: zhangyp@fudan.edu.cn

**Abstract:** In this paper, the possibility of applying different welding strategies to overlay an FeCrAl layer against corrosion from heavy liquid metal on a plain plate made of 316L austenitic stainless steel was investigated. This technology could be used in manufacturing the main vessel of CiADS, which may be considered as a more economic and feasible solution than production with the corrosion-resistant FeCrAl alloy directly. The main operational parameters of the laser welding process, including laser power, weld wire feeding speed, diameter of the welding wire, etc., were adjusted correspondingly to the optimized mechanical properties of the welded plate. After performing the standard nuclear-grade bending tests, it can be preliminarily confirmed that the low-power pulse laser with specific operational parameters and an enhanced cooling strategy will be suitable to surface an Fe-10Cr-4Al-RE layer with a thickness of approximately 1 mm on a 40 mm-thick 316L stainless steel plate, thanks to the upgraded mechanical properties incurred by refined grains with a maximum size of around 300 μm in the welded layer.

**Keywords:** CiADSvessel; FeCrAl weld overlay; austenitic stainless steel plate; mechanical tests

**Citation:** Xi, B.; Liu, B.; Li, S.; Wang, D.; Zhang, Y.; Szakálos, P.; Ejenstam, J.; Wallenius, J.; He, G.; Zhang, W. Influence of TIG and Laser Welding Processes of Fe-10Cr-4Al-RE Alloy Cracks Overlayed on 316L Steel Plate. *Materials* **2022**, *15*, 3541. https://doi.org/10.3390/ma15103541

Academic Editor: Francisca G. Caballero

Received: 9 March 2022
Accepted: 11 May 2022
Published: 15 May 2022



## 1. Introduction

Pb-based alloys are eutectics or alloys formed by adding lead as the basic material and other elements as supplementaries. The melting point of Pb-based alloys is lower than that of lead, and other physical properties are similar to those of pure Pb. Lead–bismuth eutectic (LBE) is the most commonly used alloy in the field of nuclear power, in the form of fission reactors [1]. The mass ratio between Pb and Bi in LBE is 44.5:55.5, which provides the lowest melting point among all lead–bismuth alloys [2].

The main advantages of Pb-based materials as coolants are the following: (1) due to the relatively high melting point of Pb-based materials, the Pb-based reactor could operate at normal pressure, leading to a much thinner vessel wall [3] and lower possibility of the loss of coolant accident (LOCA); (2) due to the high boiling point and high thermal conductivity of Pb-based coolant materials, Pb-based reactors can work at relatively high temperatures, which may greatly improve the system's energy efficiency; (3) due to the relatively small cross-sections for the neutron capture and scattering of Pb-based materials, a harder neutron energy spectrum could be produced in Pb-based reactors, which may improve the neutron economy and, therefore, leave extra neutrons for potential transmutation or breeding; (4) thanks to the high density of LBE, significant natural convection could be expected after losing the driving force from primary pumps [4]; (5) due to the good chemical inertness of LBE, there will be no violent chemical reactions during contact with water and air; (6) Pb-based materials can form compounds with volatile radioactive iodine and cesium nuclides, which may reduce the radioactive inventory in reactors.

In addition to the common characteristics stated above, different types of Pb-based materials, such as lead, lead–bismuth, and lead–lithium, have their own characteristics and were proposed for different types of reactors. LBE was chosen as the primary coolant material of the China-initiated Accelerator-Driven System (CiADS) [5], thanks to its low melting temperature, high boiling temperature, chemical inertness and neutron transparency. However, its corrosive property due to the dissolution of elements was reported to cause significant damage to the micro-structure of the alloy, especially at a high temperature and under severe neutron irradiation [6–8]. The corrosion effect from heavy liquid metal on stainless steel was reported to be highly affected by the oxygen content in the heavy liquid metal. Proper setting of the oxygen content in liquid is, therefore, able to permit the rapid formation of a protective oxide layer on the alloy surface, while also permitting less oxide impurity production. Generally, the introduction of aluminum into ferritic alloys may lead to better resistance to potential heavy liquid metal corrosion, thanks to the formation of protective alumina [9–14]. Due to their excellent properties, such as low cost, heat resistance and corrosion resistance, FeCrAl alloys have a wide range of applications under conditions of high temperature and they require a high level of corrosion resistance [15,16].

As the first barrier in a nuclear fuel, the choice of an accident tolerant fuel (ATF) is a very important issue in the field of nuclear fuel development [17]. After the Fukushima accident, there has been an urgent need to develop suitable fuel cladding materials to meet the performance requirements under severe accident conditions, which will, thereby, greatly improve the accident tolerance capabilities of future nuclear power plants [18–20]. Advanced cladding materials that meet these requirements can provide stronger oxidation resistance and better physical properties during severe accidents, such as SBO, LOCA, etc. [21]. This advanced cladding material must first be able to achieve the same performance as the general cladding material under the normal working condition.

In recent years, in order to solve the problem of selecting ATF cladding materials, many related studies were carried out on some candidate cladding materials. Qiu et al. [12] has made some research progress in several cladding candidate materials, such as SiC, FeCrAl and the MAX phase material $Ti_3SiC_2$. They also evaluated the physical properties of these cladding candidates in terms of thermal hydraulics and mechanical mechanics, aiming to better understand and simulate the behavior of these cladding materials.

FeCrAl alloy has the characteristics of high temperature resistance and corrosion resistance, so it has become one of the best candidate materials for ATF. Wu et al. [13] used the Monte Carlo calculation tool Serpent and fuel performance program BISON from INL to perform neutronics and fuel performance analyses of ATF. The research results show that the application of FeCrAl cladding material can significantly improve the cladding performance, such as flattening the axial temperature distribution and delaying the gap closure [17]. Additionally, hydrogen permeability was obtained for several FeCrAl alloys using a static permeation platform. The results were compared, calibrated and verified with traditional 304 stainless steel. In addition, they discussed the possibility of a protective alumina layer on the surface of FeCrAl cladding as a tritium barrier. Han et al. [15] introduced Mo as a buffer layer in the cladding, which significantly reduced the interdiffusion of Zr and Fe [22]. It was also demonstrated that a 7 μm-thick FeCrAl coating could provide excellent resistance to oxidation at temperatures up to 1200 °C [23,24]. The working limit was, therefore, significantly higher than that of traditional 316L austenitic stainless steel [25–27].

Based on the research results in the above publications, we can preliminarily conclude that FeCrAl alloy materials with excellent high temperature resistance and corrosion resistance could be considered as candidate materials for Pb-based reactors.

Ferritic FeCrAl alloys are normally used as heating elements, as they areable to work at 900–1400 °C [28]. The high Cr content in these alloys was reported to cause embrittlement at around 475 °C, due to phase transformations [29–31]. Therefore, the embrittlement characteristic of FeCrAl alloy could be minimized by reducing the Cr content. However, the introduction of Cr was reported to be able to enhance the formation of a protective Al-rich oxide layer [30]. Among FeCrAl alloys with different elementary compositions or manufacturing processes, the Fe-10Cr-4Al-RE developed by Ejenstam et al. [30,31] was reported to be able to withstand heavy liquid metal corrosion at temperatures higher than 900 °C. However, the weldability and machinability of Fe-10Cr-4Al-RE on parts made of austenitic stainless steel still needs to be verified, due to the high content of Cr and Al in it.

Tungsten inert gas (TIG) welding is an inert gas-protected fusion welding process. The heat required during the welding process is provided by an intense electric arc, which will be generated between a tungsten electrode and work pieces [9]. Compared to the traditional welding techniques, the major merits of Ar-protected TIG welding are as follows:

(1) Ar gas can prevent harmful effects from some elements in air (such as oxygen, nitrogen, hydrogen, etc.), on the arc and in the molten pool. This may reduce the burning of alloy elements and, therefore, lead to a dense, spatter-free, high-purity welding joint;
(2) The arc in the Ar-protected TIG welding is stable, which may lead to the concentration of heat and, hence, ahigher arc column temperature. A higher welding efficiency, smaller heat-affected zone, lower thermal stress, less heat deformation and smaller cracks could, therefore, be expected;
(3) Argon-protected TIG welding can be applied on almost all metals, especially those that are hard to melt or easily oxidized.

Laser welding technology is a surface coating technology involving different research fields, such as physics, chemistry, optoelectronics, materials, and so on. Compared with other surface treatment technologies, including TIG welding, laser coating technology has many advantages [10], such as the following:

(1) The thermal deformation during welding is small, which is mainly because of the smaller heat input. The heat-affected zone on the substrate will be relatively small;
(2) The cooling after the welding process is fast, which may lead to finer grains in the melt zone.The micro-structure will then be relatively dense, which is one of the characteristics ofrapid solidification;
(3) The dilution speed in the melt zone is low, which may lead to a good combination of the coating layer and the substrate metallurgically;
(4) There is basically no restriction on the selection of coating materials, either metallic or non-metallic materials could be used. It is even possible to overlay a high-melting-point material on a low-melting-point metal to improve its specific properties.
(5) The thickness of the coating layer could be selected in a large region. The coating area can also be accurately selected, which will lead to lower material consumption and, thereby, a lower capital cost.
(6) The coating process can be automated, allowing special places that are usually inconvenient to handle to be processed.

In this paper, the manufacturing process of the nuclear-grade 316L stainless steel plate with a thickness of 40 mm was introduced for the first time. The plates manufactured with these procedures will be the base sample used for the following coating process. High-frequency laser welding with low laser power was then applied to melt a thin Fe-10Cr-4Al-RE layer on a plate made of austenitic 316L stainless steel. In order to enhance the welded samples' mechanical properties, aiming to bear the nuclear-grade bending tests, the welding strategy and welding parameters were optimized based on metallurgical analyses and mechanical tests performed on the welded samples.

## 2. Material and procedures

### 2.1. Preparation of the Nuclear-Grade 316L Plate

The possibility of welding a protective Fe-10Cr-4Al-RE layer on a nuclear-grade SS316Lplain plate was studied. The plain plate with a thickness of 40 mm was manufactured according to the design of the reactor vessel wall of the China-Initiated Accelerator-Driven System (CiADS). The plain plate was hot rolled from the bulk nuclear-grade 316L steel manufactured under the standard of BTXG 018-2011 and qualified to the specifications listed in Table 1.

**Table 1.** Specifications of the plain plate samples.

| Chemical Composition | | | | | | | | | |
|---|---|---|---|---|---|---|---|---|---|
| C | Si | Mn | P | S | Cr | Ni | N | Co | Mo |
| 22 | 43 | 144 | 17 | 2 | 1737 | 1227 | 600 | 4 | 233 |
| **Mechanical Characteristics** | | | | | | | | | |
| Yield Stress, MPa | Tensile Stress, MPa | Elastic Length, % | Average Hardness, HBW | Grainess, ASTM grainess grade | | | | | |
| 277 | 565 | 59 | 163.0 | 6.5–4.0 | | | | | |

The Fe-10Cr-4Al-RE weld wire was imported from Kanthal AB, which was reported to be able to tolerate at least 1760 h corrosion from static liquid Pb at 900°C, thanks to the formation of a stable alumina protective layer at the contact surface when the oxygen level in liquid Pb was properly controlled [16]. The Fe-10Cr-4Al-RE alloy was imported as weld wires with 1.2 mm diameter.

### 2.2. FeCrAl Alloy Filler Material

The process of overlaying an FeCrAl thin layer on a plain plate made of nuclear-grade 316L stainless steel was described. The weld wire was provided by Kanthal AB, and is described in Table 2.

**Table 2.** Information of the weld wire.

| Type | Diameter, mm | Batch No. | Manufacturer | Batch Weight, kg |
|---|---|---|---|---|
| Fe-10Cr-4Al-RE alloy | 1.2 | 99324/142136 | Kanthal AB/Sandvik | 15 |

The elementary composition of the weld wire was analyzed using the Leco CS600 Carbon/Sulphur analyzer and ICP 725-ES inductively coupled plasma—optical emission spectroscopy following the ASTM-A751 standard procedures. The measured elementary vector is listed in Table 3, from which we can notice that the Al and Cr contents are relatively high, to enhance the formation of a protective oxide layer at the contact surface with the high-temperature Pb-based coolant.

**Table 3.** Elementary composition of weld wire (wt.%).

| Ni | Cr | Al | C | Si | Mn |
|---|---|---|---|---|---|
| 0.10 | 10.9 | 4.68 | 0.020 | 0.28 | 0.20 |
| Fe | Ti | Cu | Nb + Ta | S | |
| Balanced | 0.21 | 0.014 | 0.38 | <0.001 | |

### 2.3. Welding Processes

The overlay process was performed using both Ar-protected TIG welding and laser welding in individual processes, in order to determine the most economic, efficient and qualified coating strategy. A Panasonic YC400TX DC power source was used in the manual TIG welding process. Since laser welding requires more precise control of both the laser power and weld wire feeding speed, a laser welding station was proposed, consisting of a control cabinet, an MFSC 3000X single-module continuous-wave (CW) fiber laser generator, a WF007A weld wire feeding machine, a FANUC mechanical arm with a laser head, and a water cooling system, as shown in Figure 1.

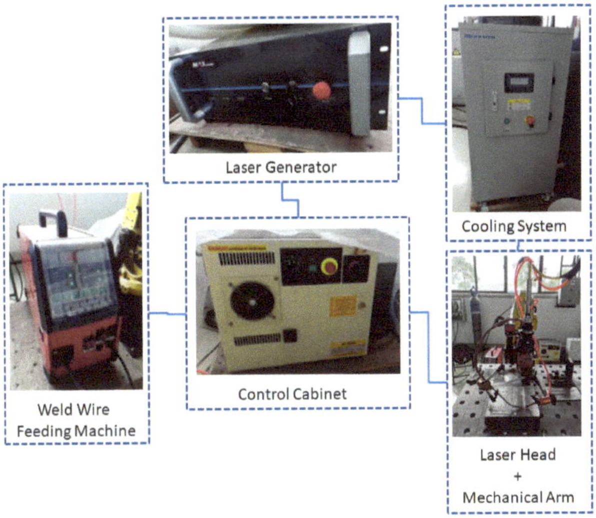


**Figure 1.** The layout of the laser welding station.

*2.4. Experimental Procedures and Results*

2.4.1. Manual TIG Welding and Results

Manual Ar-protected TIG welding was performed with the Fe-10Cr-4Al-RE weld wire in its original size, being equal to 1.2 mm in diameter. The thickness of the welding layer is around 3 mm, consisting of three sub-layers, as shown in Figure 2. The welding parameters are listed in Table 4. As shown in Figure 2, the welded surface is fairly rough and twisted, which was considered to be mainly caused by the significant melting and cooling shrinkage incurred by the high welding temperature.

**Figure 2.** Plain plate welded by manual TIG welding technique.

**Table 4.** The first-round TIG welding parameters.

| Current, A | Voltage, V | Welding Speed, mm/min | Heat Input, KJ/mm | Pre-Heating Temperature, °C |
|---|---|---|---|---|
| 130 | 12 | 85–100 | 0.936–1.101 | 150 |
| Max. interpass temperature, °C | Polarity | Tungsten bar size, mm | Protective gas | Ar gas flowrate, L/min |
| 150–165 | DCEN | 3.0 | 99.999% Ar | 13 |

Elementary compositions of the melt zone were analyzed with the OBLF QSG750 spark spectrometer with Ar gas, and are summarized in Table 5. The testing sample was polished using sandpaper with a mesh size of 80, and cleaned with high-purity alcohol. The pure Ar gas (purity of 99.999%) was supplied to the facility with apressure of 3 bar and a flowrate of 800 L/h.

**Table 5.** Elementary composition of the melt zone (wt.%).

| C | Si | Mn | P | S | Ni | Cr | Mo |
|---|---|---|---|---|---|---|---|
| 0.018 | 0.30 | 0.22 | 0.012 | 0.003 | 0.11 | 9.37 | 0.029 |
| V | Co | Cu | Ti | Nb | W | Al | / |
| 0.026 | 0.013 | 0.022 | 0.21 | 0.46 | 0.15 | 3.37 | / |

Micro-hardness tests were performed for different layers of the welded sampleusing the Wilson Tukon 2100B micro-hardness tester, and by following the standard procedure GB/T-2654.The results are listed in Table 6. We observed a reduction in hardness from the welded layer to the base material. This phenomenon was considered to be mainly caused by the difference between the elementary compositions of FeCrAl weld wire and the austenitic stainless steel base material, as can be noticed from Tables 1, 3 and 5. The hardness of the melt zone is higher than that of the base material, but lower than the FeCrAl weld wire, which was considered to be caused by the existence of Ti (formed during the hardening phase) and quenching from forced cooling.

**Table 6.** Micro-hardness in different zones of weld overlay.

| | |
|---|---|
| Parent metal area | 185, 188, 181 |
| Heat affected area | 211, 209, 196 |
| Hardfacing layer | 265, 216, 200 |

After polishing the welded surface and performing the liquid penetrate test (PT), long cracks (as circled in Figure 3) penetrating the whole welding layer can be noticed in the central part of the test plate. This confirms the limitation of applying traditional manual TIG welding and, therefore, the necessity of developing an upgraded welding strategy.

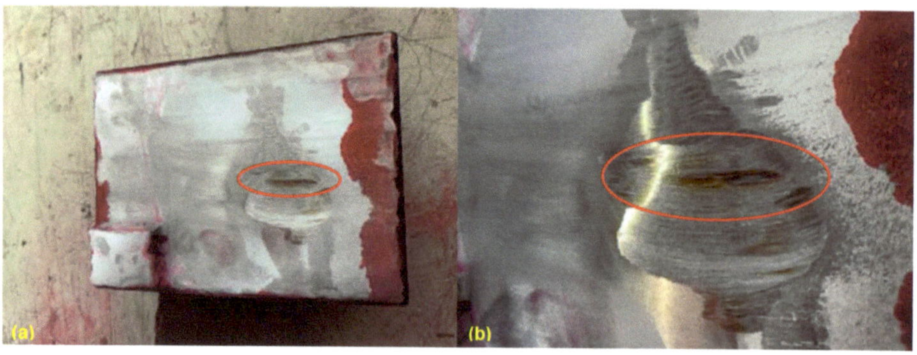


**Figure 3.** Appearance of the welded plate after liquid penetrate test (**a**) and cracks (**b**).

In addition, side bending tests were performed on the first round of TIG welded samples. Significant cracks can be observed (Figure 4), which can mainly be explained by the huge columnar grains, which cross the whole welded layer, as shown in Figure 5. The side bending tests were performed with a 30-ton bend tester (GW40F) following the ANSI/AWS B4.0-2000 standard procedures. The welded plates were then cut into test specimens with a size of 150 × 40 × 10 (mm).

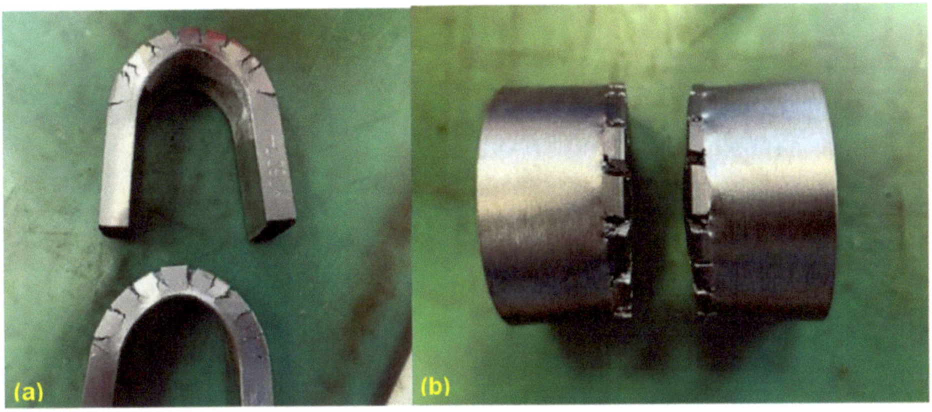


**Figure 4.** Manual TIG welded sample after side bending: side view (**a**) and top view (**b**).

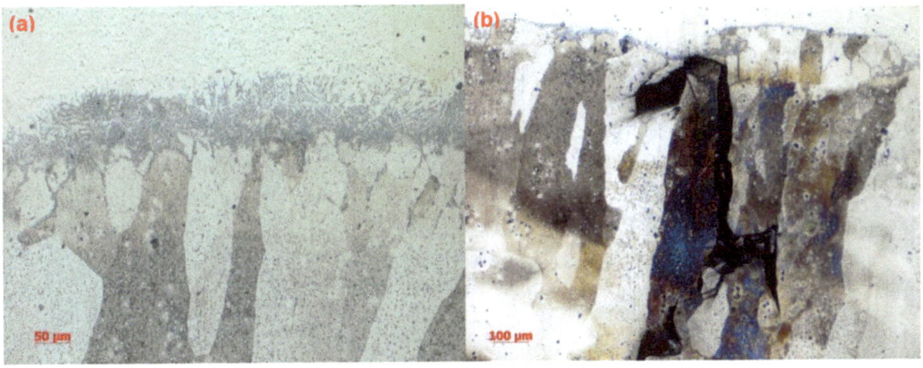


**Figure 5.** Metallographic analysis of the manual TIG welded sample: melting zone (**a**) and coating layer (**b**).

2.4.2. Laser Welding Process and Results

Since our experience of welding an FeCrAl layer on a stainless steel plate is quite limited, information from open literature was collected as a reference plan. Firstly, parameter tests were conducted to obtain a set of welding parameters that can prevent the formation of cracks during welding processes, as described above. Secondly, bending tests were conducted to check its suitability for bending.

After a series of trial tests, a proper set of welding parameters that can prevent the formation of cracks after welding processes was proposed, as listed in Table 7. Using this set of parameters, a test plate coated with Fe-10Cr-4Al-RE was manufactured, as shown in Figure 6. Side bending tests and metallographic inspections were then performed on the test plate. The test results show quite limited improvement in the mechanical behaviors of the test samples using this set of welding parameters. There are still many cracks that can be observed, as shown in Figure 7. This is mainly due to the size and the shape of grains inside the overlay layer, as can be noticed in Figure 8.

**Table 7.** Welding parameters to survive the welding processes.

| Laser Power, W | Welding Speed, mm/s | Wire Feeding Speed, mm/min | Defocusing Amount, mm | Ar Flowrate, L/min |
|---|---|---|---|---|
| 1600 | 12 | 130 | +60 | 25 |

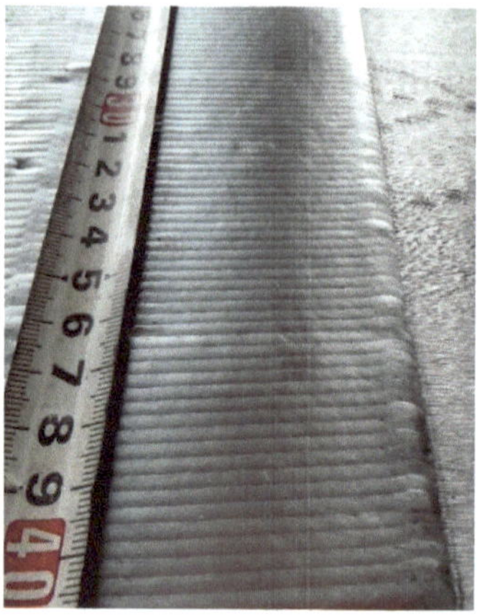

**Figure 6.** Appearance of first batch of laser weld overlay.

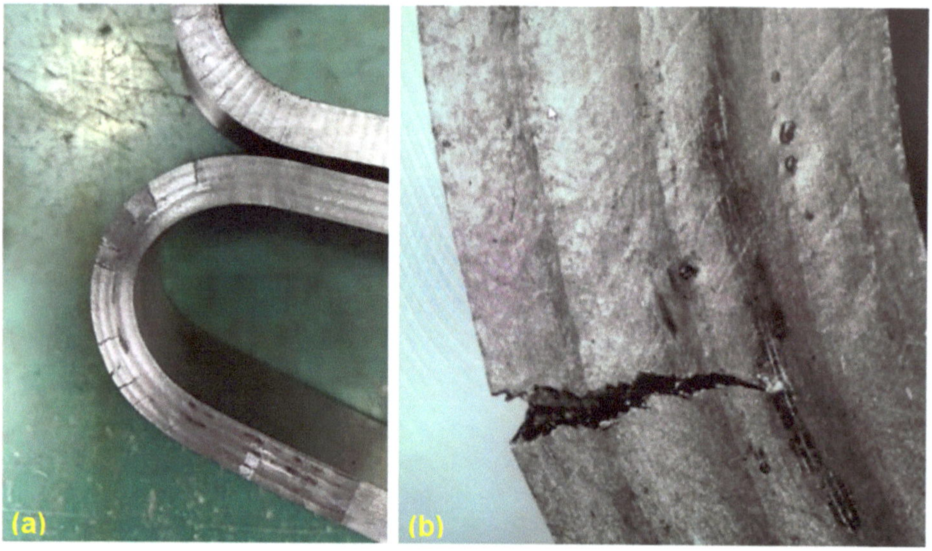


**Figure 7.** Cracks after side bending tests: test sample (**a**) and cracks (**b**).

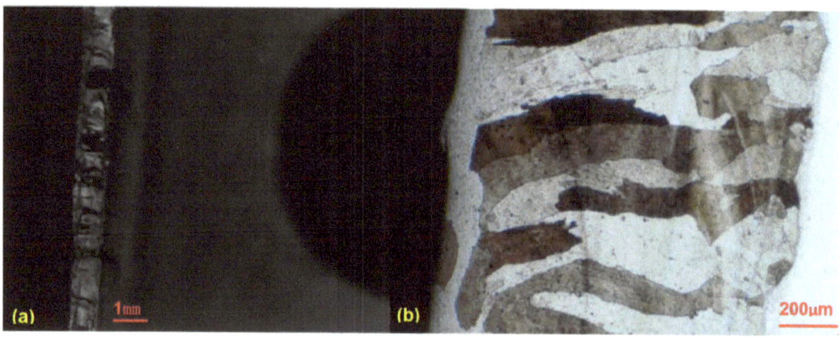


**Figure 8.** Metallographic analysis of the laser overlay plate (**a**) and the coating layer (**b**).

It was reported in this author's former articles that cracking during bending tests could be solved by reducing the grain size in welded samples, using a thinner weld wire with a 0.2 mm diameter and adjusting the welding parameters correspondingly. The laser power was set at a very low level. As discussed above, a surface layer thicker than 0.5 mm should be fully sufficient for corrosion resistance. Therefore, the thickness of the coating will not be a key problem to be considered during welding parameter optimization. The growth of crystal grains depends on the melting time at high temperature, so welding parameters can be optimized by reducing the heat input and increasing the welding speed to enhance surface cooling, and by matching the feeding speed of the weld wire to obtain a thinner coating layer with smaller grains inside.

After a series of tests, a set of optimized parameters were obtained: the laser power was set at 1500 W, the welding speed was set at 18 mm/s, the wire feeding speed was set at 100 cm/min, and the defocusing amount was set at +60 mm. The second test plate was overlaid using this set of parameters, as shown in Figure 9, and the side bending test and metallographic inspection were carried out against this sample plate, as shown in Figures 10 and 11.

**Figure 9.** Appearance of the welded plate with upgraded parameters.

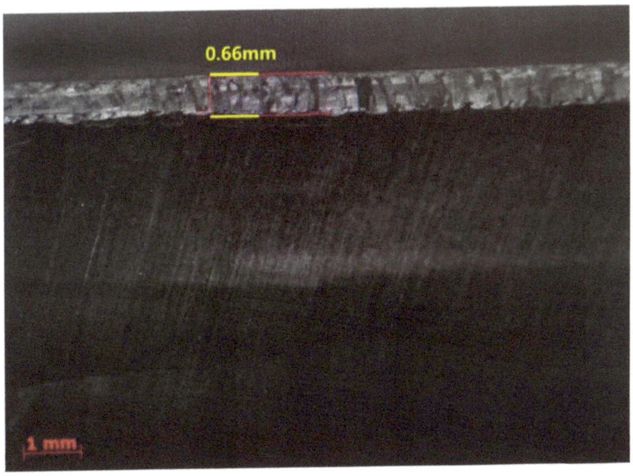


**Figure 10.** Macroscopic metallographic analysis of the welded plate with upgraded parameters.

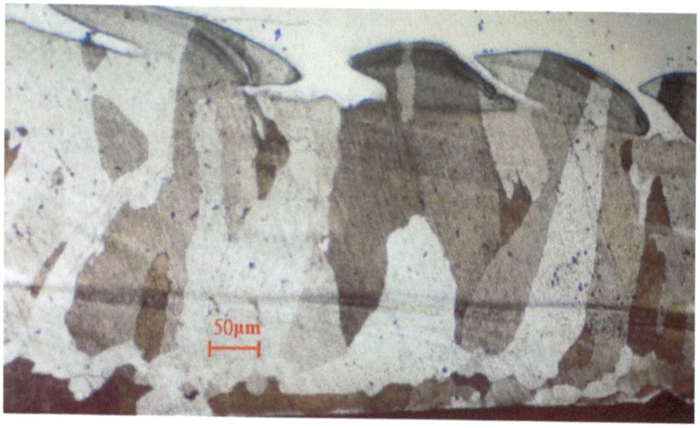


**Figure 11.** Microscopic metallographic analysis of the welded plate with upgraded parameters.

## 3. Discussion

Judging from the crack direction in the sample plate prepared with TIG welding, the source of cracks is close to the fusion line, and it will extend upward along the grain boundaries in the weld layer. Feather-like upper bainite was found in the dilution zone near these crack source points, as shown in Figure 5. The formation of cracks was then considered to be mainly caused by the brittleness of the micro-structure in the welding layer and the concentration of thermal stress. The production of upper bainite was considered to be mainly related to the chemical composition in the weld dilution zone, as well as the low interlayer temperature incurred by insufficient preheating. The welded line mainly consists of ferrite and granular brittle precipitates, which enables its plasticity and, thus, causes significant expansion.

The welding layer obtained from the laser welding using the first set of parameters is about 1.0 mm thick. The surface of this layer contained pores and fairly coarse grains, as can be observed in Figure 8. During the side bending tests, multiple cracks appeared and the crack depth exceeded more than half of the weld layer. After optimization of the laser welding parameters, the thickness of the coating layer was polished and tested. No defects could be found after PT tests, and the remaining coating layer was reported to be around 0.6–0.7 mm thick (as shown in Figure 10). As can be noticed from Figure 11, much finer columnar grains were reported to appear at the surface of the welding layer, and no obvious pores or feather-like upper bainite could be found, which could be used to explain its better mechanical properties when compared to the samples prepared using manual TIG welding. Moreover, as can be observed from Figure 12, no visible cracks were found after the side bending tests, which can preliminarily prove the feasibility of improving the mechanical performances of the sample plate by reducing the thickness of the welded layer and refining the grains inside it. This could be realized by properly adjusting the operational parameters of laser welding processes.

**Figure 12.** Result from the bending test with plunger size of 40 mm and bending angle of 180°.

## 4. Conclusions

In order to obtain a proper welding strategy to overlay an FeCrAl alloy on a nuclear-grade austenitic 316L stainless steel plate, both traditional TIG welding and high-frequency laser welding were investigated and optimized, with the aim of surviving the nuclear-grade side bending tests, to match the conditions that could possibly be incurred by temperature changes or irradiation during the operation of CiADS. After the series of laser surfacing tests performed using Fe-10Cr-4Al-RE weld wire described in this article, the following conclusions can be preliminarily obtained:

1. The flatness, straightness, and smoothness of the weld bead from the laser welding discussed in this article are far better than those of the argon arc welding stated above;
2. Upgraded welding parameters can provide refined grains in the welded layer with a size of 50–300 μm;
3. The sample plates with laser weld overlay were reported to be able to pass the nuclear-grade side bending test. This preliminarily proves the possibility of applying this set of welding parameters to overlay an Fe-10Cr-4Al-RE alloy on the inner surface of a lead-cooled fast reactor vessel;
4. Operational parameters of the laser welding process will have significant impacts on the microstructures of a welding layer, especially the grain size within it, which will affect the mechanical properties of welded samples;
5. The high-frequency low-power laser surfacing technology proposed in this article is suitable for producing a thin coating layer, comparable to that of the traditional TIG/MIG welding techniques. The thickness of a single welding layer can reach 1 mm and still be able to provide satisfying surface quality and mechanical properties.

The high-frequency low-energy laser welding system proposed in this article has been preliminarily proven to be suitable to overlay an Fe-10Cr-4Al-RE protective layer on a nuclear-grade austenitic 316L stainless steel plate, without causing cracks after welding or side bending, thanks to the refined grain structures in the melt zone. Since this kind of welded plate could be used to manufacture the wall of the CiADS main vessel, irradiation effects on the micro-structures of these plates should also be investigated by performing irradiation tests, and such tests are already planned by the authors.

**Author Contributions:** Conceptualization, B.X. and B.L.; methodology, S.L.; software, D.W.; validation, Y.Z., P.S. and J.E.; formal analysis, G.H.; investigation, W.Z.; resources, B.X.; data curation, B.L.; writing—original draft preparation, B.X.; writing—review and editing, Y.Z.; visualization, J.W.; supervision, Y.Z.; project administration, J.W.; funding acquisition, Y.Z. All authors have read and agreed to the published version of the manuscript.

**Funding:** This work is financially supported by the National Key Research and Development Project of the People's Republic of China with the project number 2020YFB1902100, and also by Shanghai Municipal Commission of Economy and Informatization with the project number GYQJ-2018-2-02.

**Institutional Review Board Statement:** Not applicable.

**Informed Consent Statement:** Not applicable.

**Data Availability Statement:** Data sharing not applicable.

**Acknowledgments:** The NEXTOWER project, part of the European program Horizon 2020, is greatly acknowledged for enabling us to study the Fe-10Cr-4Al-RE welding wire.

**Conflicts of Interest:** The authors declare no conflict of interest.

## References


1. Xu, H.; He, Y.; Luo, P. China's Accelerator Driven Sub-critical System (ADS) Program. *AAPPS Bull.* **2015**, *25*, 30–35.
2. Zhang, J. A review of steel corrosion by liquid lead and lead-bismuth. *Corros. Sci.* **2009**, *51*, 1207–1227. [CrossRef]
3. Zhang, H.; Xi, B.; Gu, L.; Li, j.; Zhang, L.; Jiang, W.; Tang, Y. R&D and challenges of key technologies for lead-based fast reactors based on the Accelerator-Driven Transmutation Research Device (CiADS). In Proceedings of the 2019 Academic Annual Meeting of the Chinese Nuclear Society, Kunming, China, 19–21 September 2019.
4. Wu, Y.; Bai, Y.; Song, Y. Study on conceptual design of lead-based research reactors in China. *Nucl. Sci. Eng.* **2014**, *34*, 201–208.
5. Wu, Y. Research progress and application prospects of lead-based reactors. *Mod. Phys. Knowl.* **2018**, *30*, 35–39.
6. Wu, Y. Design and R&D progress of ADS lead-based reactors in China. *Engineering* **2016**, *2*, 124–131.
7. Zhao, X.; Zeng, X.; Zhang, Y.; Yan, Q.; Chen, Y.; Yin, Z. Research progress in environmental corrosion of ODS steel with heavy metals. *Chin. J. Mech. Eng.* **2020**, *56*, 10. [CrossRef]
8. Zhang, Z.; Niu, Y.; Zhang, X. The role of the third component of chromium in iron-chromium-aluminum alloy. *J. Iron SteelRes.* **2007**, *19*, 5.
9. Gussev, M.N.; Field, K.G.; Yamamoto, Y. Design, properties, and weldability of advanced oxidation-resistant FeCrAl alloys. *Mater. Des.* **2017**, *129*, 227–238. [CrossRef]
10. Zhou, J.; Qiu, S.; Du, P.; Sun, Y.; Wang, H. Research progress of FeCrAl stainless steel for accident-resistant fuel cladding. *Mater. Rev.* **2017**, *31*, 47–51.
11. He, Y. *Basic Research on Solidification Characteristics and High-Temperature Mechanical Properties of High-Alumina FeCrAl Stainless Steel*; University of Science and Technology Beijing: Beijing, China, 2019.
12. Qiu, B.; Wang, J.; Deng, Y.; Wang, M.; Wu, Y.; Qiu, S. A review on thermohydraulic and mechanical-physical properties of SiC, FeCrAl and $Ti_3SiC_2$ for ATF cladding. *Nucl. Eng. Technol.* **2019**, *52*, 1–13. [CrossRef]
13. Wu, X.; Kozlowski, T.; Hales, J.D. Neutronics and fuel performance evaluation of accident tolerant FeCrAl cladding under normal operation conditions. *Ann. Nucl. Energy* **2015**, *85*, 763–775. [CrossRef]
14. Hu, X.; Terrani, K.A.; Wirth, B.D.; Snead, L.L. Hydrogen permeation in FeCrAl alloys for LWR cladding application. *J. Nucl. Mater.* **2015**, *461*, 282–291. [CrossRef]
15. Han, X.; Wang, Y.; Peng, S.; Zhang, H. Oxidation behavior of FeCrAl coated Zry-4 under high temperature steam environment. *Corros. Sci.* **2019**, *149*, 45–53. [CrossRef]
16. Zhou, Y.; Zuo, X.Q.; Sun, J.H.; Mei, J.; Sun, J.L. Effects of sintering parameters on the structures of Fe–Cr–Al extruded honeycombs. *Mater. Sci. Eng. A* **2007**, *457*, 329–333. [CrossRef]
17. Koo, B.U.; Yi, Y.; Lee, M.; Kim, B.K. Effects of particle size and forming pressure on pore properties of Fe-Cr-Al porous metal by pressureless sintering. *Met. Mater. Int.* **2017**, *23*, 336–340. [CrossRef]
18. Zhang, J.; Li, N. Review of the studies on fundamental issues in LBE corrosion. *J. Nucl. Mater.* **2008**, *373*, 351–377. [CrossRef]
19. Nuclear Energy Agency (NEA). *Handbook on Lead-Bismuth Eutectic Alloy and Lead Properties, Materials Compatibility, Thermal-Hydraulics and Technologies*; Nuclear Science; NEA: Vienna, Austria, 2015.
20. Asher, R.C.; Davies, D.; Beetham, S.A. Some observations on the compatibility ofstructural materials with molten lead. *Corros. Sci.* **1977**, *17*, 545–557. [CrossRef]
21. Müller, G.; Heinzel, A.; Konys, J.; Schumacher, G.; Weisenburger, A.; Zimmermann, F.; Engelko, V.; Rusanov, A.; Markov, V. Results of steel corrosion tests in flowing liquid Pb/Bi at 420–600 °C after 2000 h. *J. Nucl. Mater.* **2002**, *301*, 40–46. [CrossRef]
22. Heinzel, A.; Kondo, M.; Takahashi, M. Corrosion of steels with surface treatment and Al-alloying by GESA exposed in lead-bismuth. *J. Nucl. Mater.* **2006**, *350*, 264–270. [CrossRef]
23. Weisenburger, A.; Heinzel, A.; Müller, G.; Muscher, H.; Rousanov, A. T91 cladding tubes with and without modified FeCrAlY coatings exposed in LBE at different flow, stress and temperature conditions. *J. Nucl. Mater.* **2008**, *376*, 274–281. [CrossRef]
24. Weisenburger, A.; Schroer, C.; Jianu, A.; Heinzel, A.; Konys, J.; Steiner, H.; Müller, G.; Fazio, C.; Gessi, A.; Babayan, S.; et al. Long term corrosion on T91 and AISI1 316L steel in flowing lead alloy and corrosion protection barrier development: Experiments and models. *J. Nucl. Mater.* **2011**, *415*, 260–269. [CrossRef]

25. Fetzer, R.; Weisenburger, A.; Jianu, A.; Mueller, G. Oxide scale formation of modified FeCrAl coatings exposed to liquid lead. *Corros. Sci.* **2012**, *55*, 213–218. [CrossRef]
26. Jönsson, B.; Berglund, R.; Magnusson, J.; Henning, P.; Hättestrand, M. High temperature properties of a new powder metallurgical FeCrAl alloy. *Mater. Sci. Forum* **2004**, *461–464*, 455–462. [CrossRef]
27. Miller, M.; Hyde, J.; Hetherington, M.; Cerezo, A.; Smith, G.; Elliott, C. Spinodal decomposition in Fe-Cr alloys: Experimental study at the atomic level and comparison with computer models-I. Introduction and methodology. *Acta Metall. Mater.* **1995**, *43*, 3385–3401. [CrossRef]
28. Hyde, J.; Miller, M.; Hetherington, M.; Cerezo, A.; Smith, G.; Elliott, C. Spinodal decomposition in Fe-Cr alloys: Experimental study at the atomic level and comparison with computer models-II. Development of domain size and composition amplitude. *Acta Metall. Mater.* **1995**, *43*, 3403–3413. [CrossRef]
29. Zhang, Z.; Gesmundo, F.; Hou, P.; Niu, Y. Criteria for the formation of protective $Al_2O_3$ scales on Fe-Al and Fe-Cr-Al alloys. *Corros. Sci.* **2006**, *48*, 741–765. [CrossRef]
30. Dömstedt, P.; Lundberg, M.; Szakálos, P. Corrosion studies of a low alloyed Fe–10Cr–4Al steel exposed in liquid Pb at very high temperatures. *J. Nucl. Mater.* **2020**, *531*, 152022. [CrossRef]
31. Ejenstam, J.; Halvarsson, M.; Weidow, J.; Jönsson, B.; Szakalos, P. Oxidation studies of Fe10CrAl-RE alloys exposed to Pb at 550 °C for 10,000 h. *J. Nucl. Mater.* **2013**, *443*, 161–170. [CrossRef]

*materials*

Article

# Two-Steps Method to Prepare Multilayer Sandwich Structure Carbon Fiber Composite with Thermal and Electrical Anisotropy and Electromagnetic Interference Shielding

**Chuanqi Zhang [1], Lansen Bi [1], Song Shi [1], Huanhuan Wang [1], Da Zhang [1], Yan He [1,2,3,*] and Wei Li [4,*]**

1 College of Electromechanical Engineering, Qingdao University of Science and Technology, Qingdao 266061, China
2 Shandong Laboratory for Preparation and Application of High-Performance Carbon Materials, Qingdao 266061, China
3 Shandong Collaborative Innovation Center of Intelligent Green Manufacturing Technology and Equipment, Qingdao 266061, China
4 Department of Energy Engineering, Zhejiang University, Hangzhou 310027, China
* Correspondence: heyan@qust.edu.cn (Y.H.); weili96@zju.edu.cn (W.L.)

**Abstract:** Carbon fiber (CF) composites performance enhancement is a research hotspot at present. In this work, first, a sandwich structure composite, CF@(carbon nanotube/$Fe_3O_4$)/epoxy (CF@(CNT/$Fe_3O_4$)/EP), is prepared by the free arc dispersion-CFs surface spraying-rolling process method, herein, CFs in the middle layer and (CNT/$Fe_3O_4$)/EP as top and substrate layer. Then, CF@(CNT/$Fe_3O_4$)/EP (on both sides) and CFs (in the middle) are overlapped by structure design, forming a multilayer CF@(CNT/$Fe_3O_4$)/EP-CFs composite with a CFs core sheath. A small amount of CNT/$Fe_3O_4$ is consumed, (CNT/$Fe_3O_4$)/EP and CFs core sheath realize thermal and electrical anisotropy and directional enhancement, and multilayer sandwich structure makes the electromagnetic interference (EMI) shielding performance better strengthened by multiple absorption–reflection/penetration–reabsorption. From CF-0 to CF-8, CNT/$Fe_3O_4$ content only increases by 0.045 wt%, axial thermal conductivity ($\lambda_{\parallel}$) increases from 0.59 W/(m·K) to 1.1 W/(m·K), growth rate is 86%, radial thermal conductivity ($\lambda_{\perp}$) only increases by 0.05 W/(m·K), the maximum $\lambda_{\parallel}/\lambda_{\perp}$ is 2.9, axial electrical conductivity ($\sigma_{\parallel}$) increases from 6.2 S/cm to 7.7 S/cm, growth rate is 24%, radial electrical conductivity ($\sigma_{\perp}$) only increases by $0.7 \times 10^{-4}$ S/cm, the total EMI shielding effectiveness (EMI $SE_T$) increases by 196%, from 10.3 dB to 30.5 dB. This provides a new idea for enhancing CFs composite properties.

**Keywords:** carbon nanotubes; structural composites; plasma spraying

**Citation:** Zhang, C.; Bi, L.; Shi, S.; Wang, H.; Zhang, D.; He, Y.; Li, W. Two-Steps Method to Prepare Multilayer Sandwich Structure Carbon Fiber Composite with Thermal and Electrical Anisotropy and Electromagnetic Interference Shielding. *Materials* **2023**, *16*, 680. https://doi.org/10.3390/ma16020680

Academic Editor: Alexander N. Obraztsov

Received: 9 December 2022
Revised: 30 December 2022
Accepted: 5 January 2023
Published: 10 January 2023



## 1. Introduction

With the rapid development of aerospace, transportation, energy, medical and health fields, there is an urgent need for materials with excellent thermal/electrical conductivity properties and electromagnetic interference shielding effectiveness (EMI SE) to adapt to the work in complex environments. Carbon fiber (CF) composites are widely used because of its high strength, high modulus, light weight and easy molding [1–5]. However, the poor magnetic property for CFs limits the further improvement of EMI SE [6]. In addition, the epoxy resin (EP), which is often used as the matrix of CF composites, has advantages of light weight, designability and easy processing [7], but its low intrinsic thermal conductivity (0.1~0.4 W/(M·K)) [8,9] and EMI SE (about 2 dB) [10] limit the performance of CF composites. Therefore, the preparation of CF composites with excellent thermal/electrical conductivity properties and EMI SE has become a research hotspot.

Adding nanofillers is one of the effective methods to prepare high performance composites [11–13]. Carbon nanotubes (CNTs) have excellent thermal and electrical properties [14,15] and are often used as an ideal material to enhance thermal/electrical conductivity properties of CF composites [16,17]. Moreover, because of their good dielectric loss characteristic, CNTs are a good choice for EMI shielding materials [18–20], and CFs have similar characteristic [21–23]. However, if CNTs and CFs are simply combined, although CF-CNT composites have good thermal and electrical properties, lacking magnetic property and impedance mismatch [6] will make CF-CNT composites have weak electromagnetic wave absorption and high electromagnetic reflectance, which result in secondary electromagnetic pollution [24,25]. The matching of electrical and magnetic properties is the key to obtain good shielding effect in the wide frequency range [26]. Therefore, in order to supplement the magnetic property lacking for CFs and CNTs and enhance magnetic loss, $Fe_3O_4$, $Fe_2O_3$, Fe, Ni, Co and other magnetic particles are usually introduced to cooperate with CNTs [27–30]. Main methods include in situ growth [31,32] and mechanical blending [33,34]. Among of them, $Fe_3O_4$ has a low toxicity and good biocompatibility as a more efficient shielding material. Furthermore, because of the large saturation magnetization of $Fe_3O_4$, they can provide a high value of complex permeability. $Fe_3O_4$ can exhibit the skin effect, their high resistivity allowing the electromagnetic waves to enter effectively. Therefore, CNT/$Fe_3O_4$ as a material with dual magnetic and dielectric properties could be important to achieve excellent thermal, electrical and EMI shielding effectiveness. On this basis, Li et al. [35,36] propose a free arc dispersion method, which can rapidly disperse nanomaterials and produce nanomaterials dispersion fog with good dispersion degree in the air. In addition, this method can disperse variety of nanomaterials at the same time, and the dispersed nanomaterials have a good mixing effect. This seems to well meet the need of collaborative use for the $Fe_3O_4$ and CNTs.

Another method to prepare high-performance composites is structure design to obtain composites with specific functions [37–43]. On the one hand, in order to meet the requirements of directional heat dissipation or electrical conductivity for composites, composites are required to have anisotropy [37,38]. On the other hand, the single-layer shielding structure is not easy to achieve high absorption loss, so sandwich structure, multilayer structure and porous structure begin to appear [39–42]. Based on the difference between the radial and axial thermoelectric properties of CFs and CNTs, if CNTs are extended along the CFs axial direction to form a CNTs network that is attached to the CFs surface, meanwhile, the $Fe_3O_4$ is mixed in the CNTs network by the free arc dispersion method, and CF composites with both thermal and electrical anisotropy and the EMI shielding property can be obtained. Furthermore, multilayer CF composites constructed by the above material can not only achieve the directional enhancement of thermal and electrical anisotropy for CF composites but also realize EMI shielding performance enhancement by multiple absorption–reflection/penetration–reabsorption when electromagnetic waves pass through each layer of CF composites.

Here, as step one, CF@(CNT/$Fe_3O_4$)/EP with sandwich structure is prepared by the free arc dispersion-CFs surface spraying-rolling process method, CFs in the middle layer, and (CNT/$Fe_3O_4$)/EP as top and substrate layer. Step two, CF@(CNT/$Fe_3O_4$)/EP (on both sides) and CFs (in the middle) are overlapped by structure design, forming multilayer CF@(CNT/$Fe_3O_4$)/EP-CFs composite with CFs core sheath. The structural morphology of CNT/$Fe_3O_4$ and CF@(CNT/$Fe_3O_4$)/EP are characterized by scanning electron microscopy (SEM), Raman spectroscopy and X-ray diffraction (XRD). The influence of multilayer sandwich structure on thermal and electrical anisotropy and EMI SE of multilayer CF@(CNT/$Fe_3O_4$)/EP-CFs composite is studied.

## 2. Materials and Experiments

### *2.1. Materials*

CNTs (GT-300, length 15–30 μm, diameter 5–15 nm) were provided by Shandong Dazhan Nano Materials Co., Ltd., Binzhou, China. $Fe_3O_4$ (diameter 20 nm) was purchased from Nanjing Emperor Nano Materials Co., Ltd., Nanjing, China. CF (T700SC, diameter 7 μm) was supplied from Lianyungang Zhongfu Shenying Carbon Fiber Co., Ltd., Lianyungang, China. Epoxy resin (MF-4101H) and curing agent (ZH-520), Curing temperature $T_1$ = 150 °C, $T_2$ = 180 °C, curing time $T_1$ = 2 h, 2 = 2 h, was obtained from

Hubei Zhen Zhengfeng Advanced Materials Co., Ltd., Huanggang, China. Deionized water (DI water) was used as a dispersive working medium.

### 2.2. Experiments

Step1, free arc dispersion-CFs surface spraying-rolling process method

CNTs, $Fe_3O_4$ and DI water were mixed at a mass ratio of 1:3:10 and thoroughly stirred for 10 min, putting in the mold and applying 10 kg pressure to extrude into a cylindrical block (diameter 30 mm and height 10 mm). According to the free arc dispersion method of Li et al. [35,36], the cylindrical block was placed between the high-voltage pulse electrodes for dispersion, the voltage was 12 KV, the frequency was 10 Hz, the positive electrode used titanium grid, the negative electrode used titanium plate and CNT/$Fe_3O_4$ dispersion fog was obtained. At the same time, CNT/$Fe_3O_4$ dispersion fog passed through the spraying channel and was sprayed on continuously moving CFs surface by negative pressure airflow traction, and CFs movement speed was 0.01 m/s. The sprayed CFs moved into the heating box for heating at 100 °C to obtain CF@(CNT/$Fe_3O_4$). Finally, EP was poured on CF@(CNT/$Fe_3O_4$), after rolling, CNT/$Fe_3O_4$ was laid on the CFs surface to construct CNT/$Fe_3O_4$ network and CF@(CNT/$Fe_3O_4$)/EP was prepared. The above processes were simultaneous and continuous.

Step2, structure design

Pure CFs was placed in the middle as the core sheath, and CF@(CNT/$Fe_3O_4$)/EP was overlapped on the upper and lower sides of pure CFs, putting into the mold, and transferring to the heating box for curing, multilayer CF@(CNT/$Fe_3O_4$)/EP-CFs composite was obtained. Curing temperature $T_1$ = 150 °C, $T_2$ = 180 °C, curing time $T_1$ = 2 h, $T_2$ = 2 h. The sample size was 2 mm × 12 mm × 20 mm.

After calculation, the CFs volume fraction was 60% and the volume fraction of EP was 40% in the sample. Considering the sample performance gradient and consistency of composite size, the total number of layers for CF@(CNT/$Fe_3O_4$)/EP and pure CFs was fixed to 8. The schematic diagram of CFs overlapping method and the description of treatment for each experimental group were shown in Table 1.

**Table 1.** Description of each experimental group treatment.

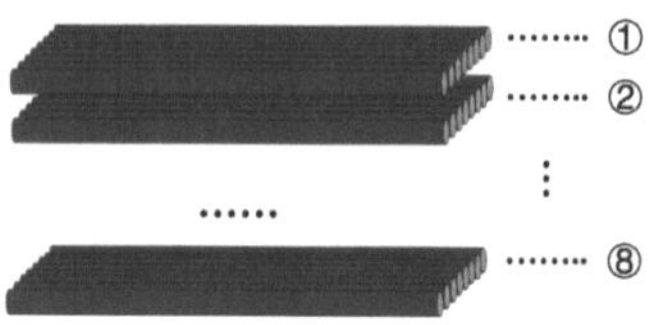


| Experiment Group | Spraying and Roll Treatment | Filler | The Amount of CNT + $Fe_3O_4$ (mg/cm$^3$) | The Content of (CNT/$Fe_3O_4$) |
|---|---|---|---|---|
| CF-0 | None | None | 0 | 0 |
| CF-2 | ①⑧ | | 0.04 + 0.12 = 0.16 | 0.011 wt% |
| CF-4 | ①②⑦⑧ | CNT | 0.08 + 0.24 = 0.32 | 0.023 wt% |
| CF-6 | ①②③⑥⑦⑧ | and | 0.12 + 0.36 = 0.48 | 0.034 wt% |
| CF-7 | ①②③⑤⑥⑦⑧ | $Fe_3O_4$ | 0.14 + 0.42 = 0.56 | 0.040 wt% |
| CF-8 | ①②③④⑤⑥⑦⑧ | | 0.16 + 0.48 = 0.64 | 0.045 wt% |

### 2.3. Characterizations

Field emission scanning electron microscope SEM (SU-8010, Hitachi, Tokyo, Japan) was applied to observe the surface distribution and morphology of CFs and composites. Raman spectrometer (InVia Reflex, Renishaw, London, UK) was used to analyze the material structure of CNT and $Fe_3O_4$, and the laser wavelength was 532 nm. X-ray diffractometer XRD (MiniFlex 600, Rigaku, Tokyo, Japan) was used to characterize the atomic structure of CNT and $Fe_3O_4$, and the scanning speed was 10°/min, the range was 20–80°. Thermal constant analyzer (TPS2500S, Hot Disk, Uppsala, Sweden) was used to test the thermal conductivity of multilayer CF@(CNT/$Fe_3O_4$)/EP-CFs composite according to the standard of ISO22007-2-2015. Electrical conductivity of multilayer CF@(CNT/$Fe_3O_4$)/EP-CFs composite was measured by four probes resistance tester (RTS-8, Guangzhou Four Probes Technology, Guangzhou, China), and micro-current tester (ST2643, Suzhou Jingge, Suzhou, China) was used to test interlaminar resistivity of multilayer CF@(CNT/$Fe_3O_4$)/EP-CFs composite. Vibrating sample magnetometer VSM (7404, LakeShore, OH, USA) was employed to test the magnetization hysteresis loops of CFs, CNT/$Fe_3O_4$ and CF@(CNT/$Fe_3O_4$)/EP at room temperature. Vector network analyzer (ZNB20, Rohde & Schwarz, Munich, Germany) was employed to measure the $S_{11}$, $S_{22}$, $S_{12}$ and $S_{21}$ parameters of multilayer CF@(CNT/$Fe_3O_4$)/EP-CFs composite according to the standard of ASTM D5568-08, frequency was X-band (8.2–12.4 GHz). the total EMI SE ($SE_T$), reflection EMI SE ($SE_R$) and the absorption EMI SE ($SE_A$) of multilayer CF@(CNT/$Fe_3O_4$)/EP-CFs composite were calculated according to the following formula [44]:

$$SE_T = 10\lg\left(\frac{1}{|S_{12}|^2}\right) \tag{1}$$

$$SE_R = 10\lg\left(\frac{1}{1-|S_{11}|^2}\right) \tag{2}$$

$$SE_A = 10\lg\left(\frac{1-|S_{11}|^2}{|S_{12}|^2}\right) \tag{3}$$

## 3. Results and Discussion

Figure 1 is the schematic diagram of multilayer CF@(CNT/$Fe_3O_4$)/EP-CFs composite preparation process. "Free arc dispersion" can disperse CNTs and $Fe_3O_4$ at the same time and obtain the CNT/$Fe_3O_4$ dispersion fog with well mix and dispersion degree. "CFs surface spraying" can spay CNT/$Fe_3O_4$ dispersion fog onto the CFs surface rapidly. "Rolling process" can flatten the 3D-CNT/$Fe_3O_4$ and form the 2D-(CNT/$Fe_3O_4$)/EP layer, while making CNTs has direction, which is beneficial to enhance the interlayer insulation performance, achieving thermal and electrical anisotropy of CF composites. Based on this, a sandwich structure composite (CF@(CNT/$Fe_3O_4$)/EP) with CFs in the middle layer and (CNT/$Fe_3O_4$)/EP as top and substrate layer is prepared. Furthermore, through the structure design, the CFs is placed in the middle as core sheath and CF@(CNT/$Fe_3O_4$)/EP is placed on both sides to prepare a multilayer sandwich structure CF composite (multilayer CF@(CNT/$Fe_3O_4$)/EP-CFs composite).

As is shown in Figure 2a, the left side is pure CFs, and the middle and right side are the CFs sprayed with CNT/$Fe_3O_4$ dispersion fog. It can be clearly seen that pure CFs has a light color and luster, but the CFs sprayed with CNT/$Fe_3O_4$ dispersion fog appears darker color. This is because the adsorption of CNTs and $Fe_3O_4$ on the CFs surface and changes the diffuse reflection of CFs surface. Figure 2b shows that pure CFs has a smooth surface without any substance. Figure 2c,d show the attachment of CNT/$Fe_3O_4$ when CFs moving speed is 0.01 m/s and 0.02 m/s, respectively. The faster CFs moving speed, the less CNT/$Fe_3O_4$ is deposited, and the lighter color is appeared on the macroscopic (Figure 2a, right). At the same time, CNTs and $Fe_3O_4$ have high dispersion degree and without obvious agglomeration, CNTs is connected to each other and extend to the radial and axial directions of CFs, presenting a 3D distribution, $Fe_3O_4$ is interspersed in the CNTs network, and adsorbed on the CFs surface. On the one hand, CNTs and $Fe_3O_4$ are coated by the size agent on the CFs surface, which establishes the physical association between CFs and CNT/$Fe_3O_4$ [45]. On the other hand, this may be related to the high specific surface area of CFs [46]. Figure 2e shows the sandwich structure CF@(CNT/$Fe_3O_4$)/EP composite obtained after rolling, with CFs in the middle and the thickness of (CNT/$Fe_3O_4$)/EP distributed on both sides is about 2 μm, which is uniformly attached to the CFs surface. It can be clearly seen in Figure 2f,g that CNTs and $Fe_3O_4$ are coated in EP, in which CNTs is attached to the CFs surface and distribute in the axial direction of CFs only, and $Fe_3O_4$ is interspersed in CNTs network with uniform distribution. This morphology is obviously different from Figure 2c,d; this indicates that the effect of rolling makes CNT/$Fe_3O_4$ change from 3D to 2D planar structure, which is conducive to maintaining the insulation between CFs layers.

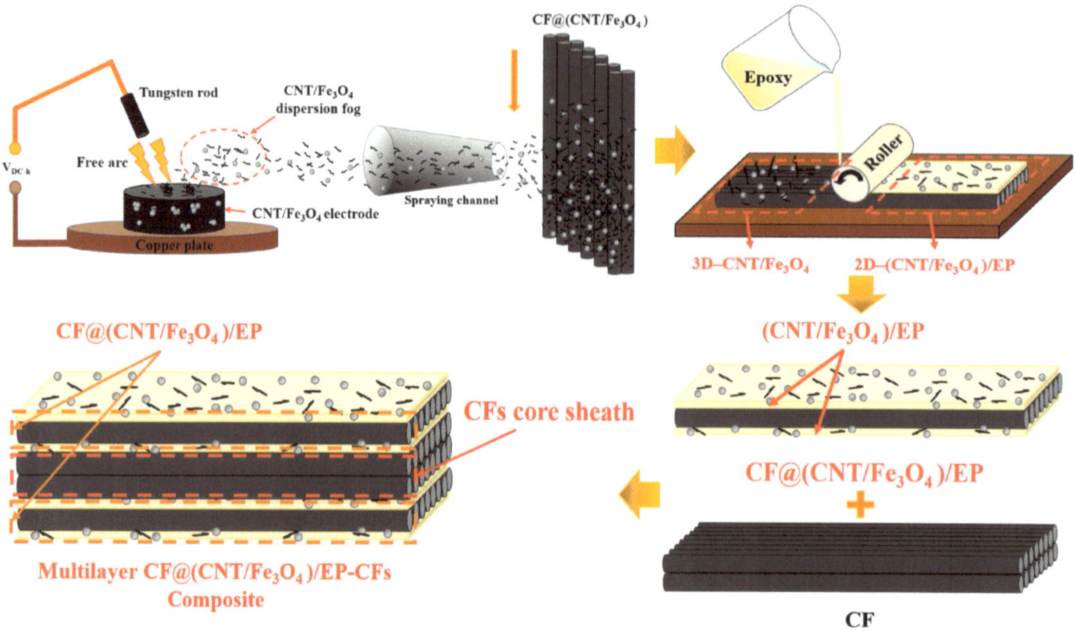


**Figure 1.** Schematic diagram of CF@(CNT/$Fe_3O_4$)/EP composite preparation process.

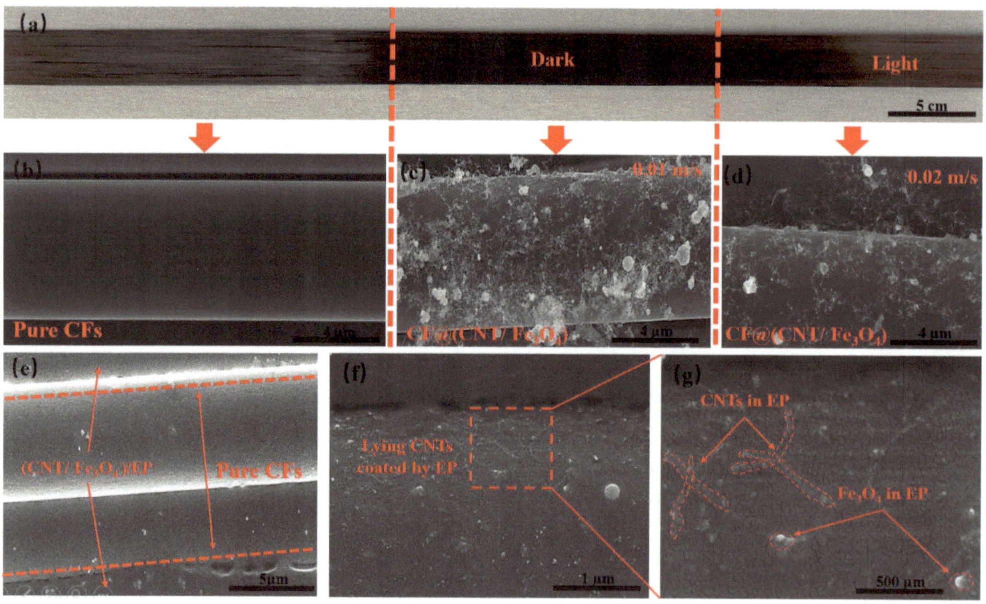


**Figure 2.** (**a**) Pure CFs at the left, CF@(CNT/$Fe_3O_4$) with high content CNT/$Fe_3O_4$ at middle, and low content CNT/$Fe_3O_4$ at right. (**b**) SEM of pure CFs. (**c**) CF@(CNT/$Fe_3O_4$) with no rolling and high content CNT/$Fe_3O_4$, and (**d**) with no rolling and low content CNT/$Fe_3O_4$. (**e**) CF@(CNT/$Fe_3O_4$)/EP with rolled treatment, CFs in the middle and CF@(CNT/$Fe_3O_4$) on both sides. (**f**,**g**) CF@(CNT/$Fe_3O_4$)/EP locally enlarged image; lying CNTs are coated by EP.

Considering that CNTs and $Fe_3O_4$ may change their properties under the action of free arc, $Fe_3O_4$ may be converted into $Fe_2O_3$ at high temperature [47]. The Raman of CNT/$Fe_3O_4$ dispersion fog obtained using the free arc dispersion method is compared with pure CNTs (Figure 3a). CNT/$Fe_3O_4$ dispersion fog has characteristic peaks at 1341 $cm^{-1}$ (D-line) and 1578 $cm^{-1}$ (G-line), which are the characteristic peaks of carbonaceous compounds [48,49]; this parameter complies with the CNT standard spectrum. In addition, the $I_D/I_G$ values of CNTs/$Fe_3O_4$ and CNTs are 1.15 and 1.12, respectively; this indicates that the graphitization degree of CNTs is not affected by the free arc. Figure 3b shows the comparison of CNT/$Fe_3O_4$ dispersion fog Raman image and $Fe_3O_4$ standard spectrum, and the result is also consistent [50,51]. It demonstrated that the structure of CNTs and $Fe_3O_4$ do not change, and CNT/$Fe_3O_4$ dispersion fog has a higher purity, only containing CNTs and $Fe_3O_4$; there are no other substances. XRD image (Figure 3c) shows that the diffraction peaks of CNTs and $Fe_3O_4$ are in good agreement with CNT/$Fe_3O_4$ dispersion fog, respectively; there are corresponding diffraction peaks at particular diffraction angles [52,53]. The above characterizations indicate that CNTs and $Fe_3O_4$ maintain good material structure during free arc dispersion and spraying.

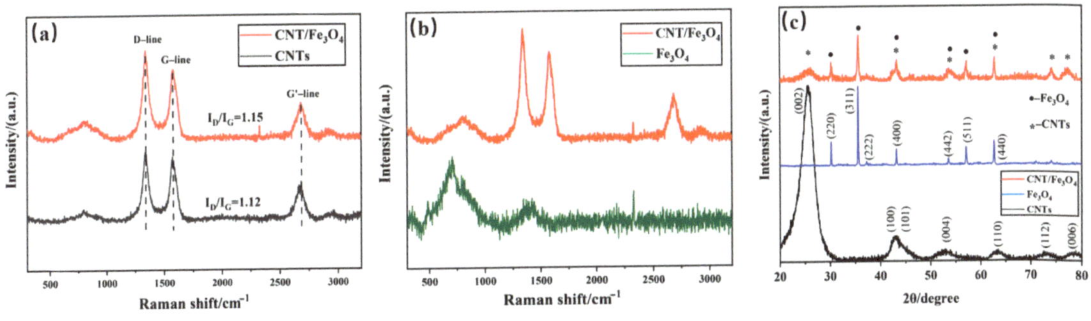


**Figure 3.** Raman spectral comparison of CNT/$Fe_3O_4$ and CNTs (**a**), and CNT/$Fe_3O_4$ and $Fe_3O_4$ (**b**), XRD of $Fe_3O_4$, CNTs and CNT/$Fe_3O_4$ (**c**).

Radial thermal conductivity ($\lambda_\perp$) of the multilayer CF@(CNT/$Fe_3O_4$)/EP-CFs composite is shown in Figure 4a. The $\lambda_\perp$ of CF-0 is 0.38 W/(m·K), and the $\lambda_\perp$ of CF-2 to CF-7 remains stable at about 0.38 W/(m·K) with the increase of CNT/$Fe_3O_4$. This is because CNT/$Fe_3O_4$ changes from 3D to 2D plane due to the rolling treatment. CNTs with excellent thermal conductivity (about 3000 W/(m·K)) [54] are attached to the CF surface and covered by EP [8] with high insulation, forming (CNT/$Fe_3O_4$)/EP. It makes cross-plane heat conduction in the multilayer CF@(CNT/$Fe_3O_4$)/EP-CFs composite not easy. In addition, due to the presence of contact thermal resistance between CFs [55], the CFs core sheath in CFs 2 to CFs 7 forms radial thermal insulation layer. Both (CNT/$Fe_3O_4$)/EP and CFs core sheath form the multilayer thermal insulation system. $\lambda_\perp$ of CF-8 increases slightly. From CF-7 to CF-8, $\lambda_\perp$ increases from 0.38 W/(m·K) to 0.44 W/(m·K). The main reason is that CF-8 does not contain a CFs core sheath; therefore, the radial thermal insulation layer is lost, but because of the (CNT/$Fe_3O_4$)/EP, the increase of $\lambda_\perp$ is not significant.

Axial thermal conductivity ($\lambda_\parallel$) of the multilayer CF@(CNT/$Fe_3O_4$)/EP-CFs composite is shown in Figure 4b. Different from $\lambda_\perp$, with the increase of CNT/$Fe_3O_4$ content, $\lambda_\parallel$ increases. When the amount of CNT/$Fe_3O_4$ is from 0 to 0.56 mg/$cm^3$ (CF-0 to CF-7), $\lambda_\parallel$ increases from 0.59 W/(m·K) to 1.1 W/(m·K), and the growth rate is 86%. The reason is that CFs has excellent thermal conductivity [56], and the axial heat conduction is not affected by EP and CFs core sheath. In addition, CNTs in the (CNT/$Fe_3O_4$)/EP forms a good thermal conductivity network [57]; the higher CNTs content, the more abundant CNTs network, and the higher heat conduction efficiency. When the amount of CNT/$Fe_3O_4$ is changed from 0.56 to 0.64 mg/$cm^3$ (CF-7 to CF-8), $\lambda_\parallel$ hardly changes, but the instability (standard deviation) increases. The main reason is that CF-8 does not have CFs core sheath, the barrier of radial heat conduction is greatly reduced, so the heat conduction has a component in radial. Thus, the increase in axial thermal conductivity is limited and the heat transfer randomness is increased.

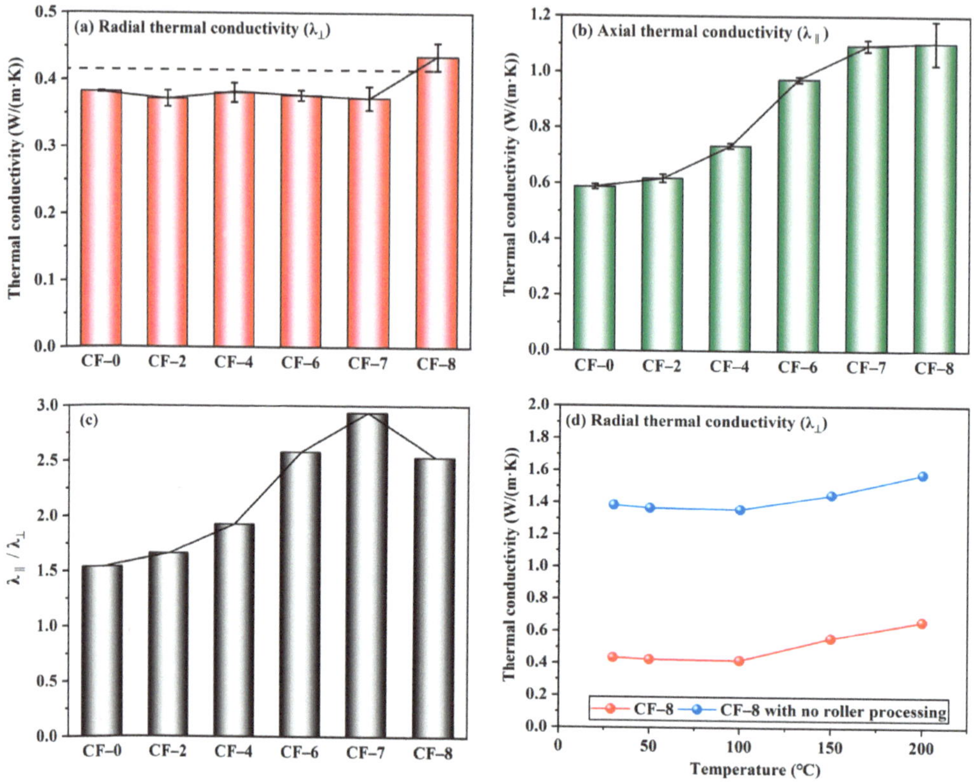


**Figure 4.** Radial thermal conductivity $\lambda_\perp$(**a**), axial thermal conductivity $\lambda_\parallel$ (**b**), and $\lambda_\parallel/\lambda_\perp$(**c**) of multilayer CF@(CNT/$Fe_3O_4$)/EP-CFs composite, $\lambda_\perp$ comparison of roll treatment (**d**).

The difference between $\lambda_\parallel$ and $\lambda_\perp$ indicates that the CFs core sheath and (CNT/$Fe_3O_4$)/EP have influence on the thermal anisotropy of multilayer CF@(CNT/$Fe_3O_4$)/EP-CFs composite. Figure 4c directly represents the difference between $\lambda_\parallel$ and $\lambda_\perp$of multilayer CF@(CNT/$Fe_3O_4$)/EP-CFs composite, the larger value of $\lambda_\parallel/\lambda_\perp$, the more significant thermal anisotropy. From CF-0 to CF-7, $\lambda_\parallel/\lambda_\perp$ gradually increases, while CF-7 to CF-8 starts to decrease. Obviously, the $\lambda_\parallel/\lambda_\perp$ of CF-7 is higher than CF-8; this is attributed to the CFs core sheath in CF-7. On the one hand, the CFs core sheath stabilizes $\lambda_\perp$; on the other hand, the CFs core sheath eliminates the radial component of heat conduction to ensure $\lambda_\parallel$ promotion. For CF-8, the large $\lambda_\perp$ and the similar $\lambda_\parallel$ make its thermal anisotropy insignificant compared to CF-7. Figure 4d shows that within the 30–200 °C, the $\lambda_\perp$ of CF-8 with rolling treatment (about 0.4 W/(m·K)) is lower than no rolling treatment (about 1.4 W/(m·K)). This shows from the performance point of view that the (CNT/$Fe_3O_4$)/EP formed by rolling can effectively reduce the heat transfer between layers. Combined with the difference between $\lambda_\parallel$ and $\lambda_\perp$, the main reason is that CNTs have high axial thermal conductivity [58], and rolling makes CNTs attach to the CFs surface and extend along the axial direction of CFs. At this time, heat can be transferred along the CFs axial direction; however, due to the coverage of EP and the direction of CNTs, radial heat transfer is difficult.

Radial electrical conductivity ($\sigma_{\perp}$) of multilayer CF@(CNT/$Fe_3O_4$)/EP-CFs composite is shown in Figure 5a. The $\sigma_{\perp}$ of CF-0 to CF-7 is generally stable, maintaining at $1.1 \times 10^{-4}$ (S/cm), while the $\sigma_{\perp}$ of CF-8 is increased to $1.7 \times 10^{-4}$ (S/cm), showing a slight improvement. The main reason is that CNTs has low resistance/high electrical conductivity ($10^5$–$10^7$ S/m) [59,60], and adding CNTs to the composite can improve electrical conductivity. Similar to the thermal conductivity, due to the insulation effect of (CNT/$Fe_3O_4$)/EP and CFs core sheath, the cross-plane electrical conduction in multilayer CF@(CNT/$Fe_3O_4$)/EP-CFs composite is difficult to carry out. Because CF-8 does not have CFs core sheath and CNTs enhance the electrical conductivity of EP [61], so the $\sigma_{\perp}$ of CF-8 obtains some improvement.

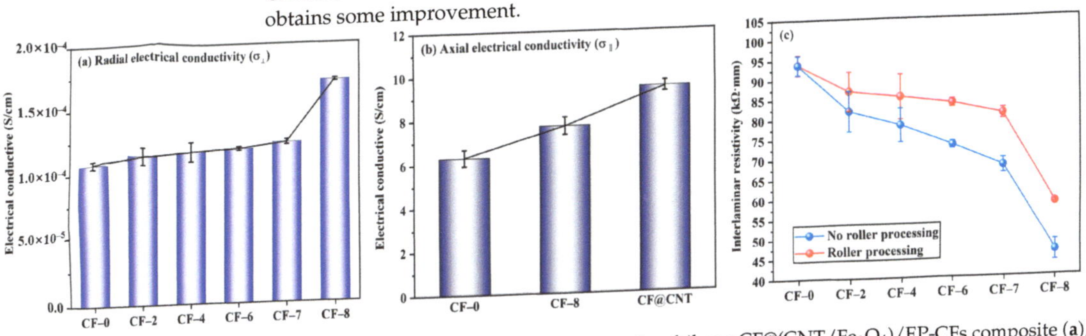


**Figure 5.** Radial electrical conductivity $\sigma_{\perp}$ of multilayer CF@(CNT/$Fe_3O_4$)/EP-CFs composite (**a**), axial electrical conductivity $\sigma_{\parallel}$comparison of CF, CF@ (CNT/$Fe_3O_4$) and CF@CNT composite (**b**), interlaminar resistivity comparison of roll treatment (**c**).

As shown in Figure 5b, axial electrical conductivity ($\sigma_{\parallel}$) is higher than $\sigma_{\perp}$, the $\sigma_{\parallel}$ of CF-0, CF-8 and CF@CNTs (the content of CNTs is 0.64 mg/cm$^3$) are 6.2 S/m, 7.7 S/m and 9.4 S/m, respectively, showing increase trend. The main reason is that $\sigma_{\parallel}$ is not restricted by (CNT/$Fe_3O_4$)/EP, CFs core sheath and interlamination contact resistance, and CFs have high axial electrical conductivity (about 670 S/cm) [62]. The $\sigma_{\parallel}$ of CF-8 is higher than CF-0 because CNTs is contained in the filler, and the CNTs direction is along the CFs axial direction, which helps to improve the axial electrical conductivity of EP and composite. In addition, although the same mass of CNTs (0.64 mg/cm$^3$) is added in CF@CNT composite, the $\sigma_{\parallel}$ of CF@CNT is higher than CF-8. This is attributed to the fact that $Fe_3O_4$ has poor electrical conductivity [63], CF-8 contains 0.48 mg/cm$^3$ $Fe_3O_4$ and CNTs content is much lower than CF@CNT, which makes low electrical conductivity for CF-8.

Figure 5c shows that the interlaminar resistivity of multilayer CF@(CNT/$Fe_3O_4$)/EP-CFs composite with rolling is generally higher than no rolling. The main reason is that CNTs no rolling may penetrate EP, thus connecting adjacent CFs, forming CFs-CNTs-CFs interlayer electric conduction pathway, which reduces the macroscopic resistivity and influences the interlamination insulation performance of the composite.

The magnetic property of CFs, CNT/$Fe_3O_4$ powder and CF@(CNT/$Fe_3O_4$)/EP are tested, and the results are shown in Figure 6a. CFs has no magnetic, and the saturation magnetization (Ms) of CNT/$Fe_3O_4$ powder is 40 emu/g, when combined with CFs and EP, the Ms decreases to 2.6 emu/g, which is mainly attributed to CNT/$Fe_3O_4$ is coated [64]. It can be seen from local magnification (Figure 6b) that the coercivity (Hc) of CNT/$Fe_3O_4$ powder and CF@(CNT/$Fe_3O_4$)/EP are 66.6Oe and 64.6Oe, respectively. The similar Hc values indicate that free arc has no effect on the antidemagnetization ability of CNT/$Fe_3O_4$.

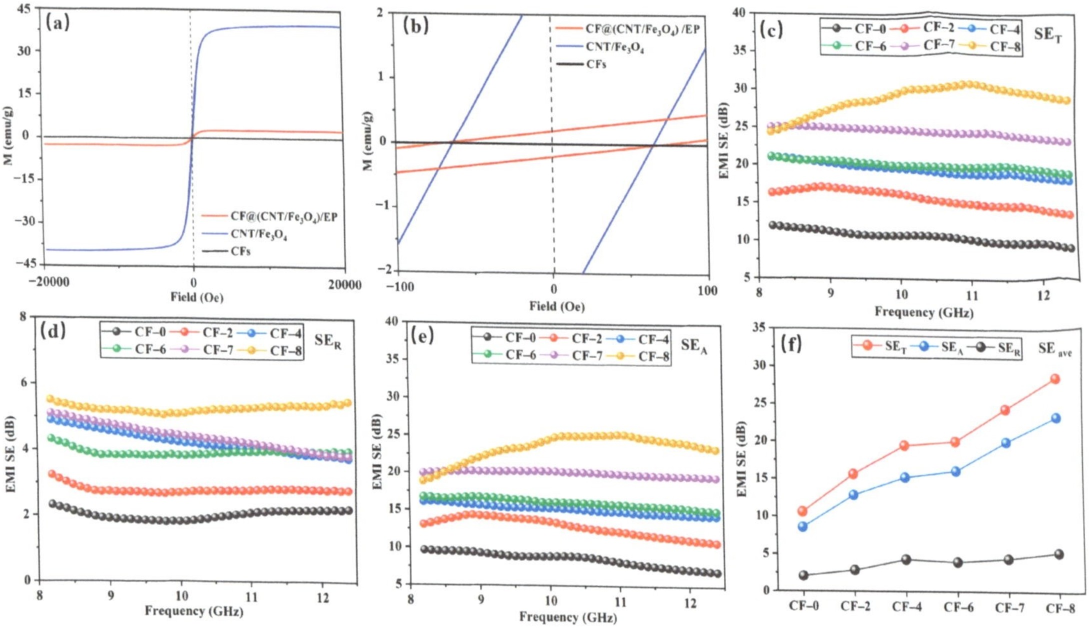

**Figure 6.** Magnetic hysteresis loops at room temperature (**a**,**b**), EMI $SE_T$ (**c**), EMI $SE_R$ (**d**), EMI $SE_A$ (**e**) and EMI $SE_{ave}$ (**f**) at 8.2 to 12.4 GHz.

The $SE_T$, $SE_R$ and $SE_A$ of CF-0 to CF-8 are shown in Figure 6c–e. The higher CNT/$Fe_3O_4$ content, the higher $SE_T$, $SE_R$ and $SE_A$ value, and the $SE_A$ value is greater than $SE_R$. In CF@(CNT/$Fe_3O_4$)/EP, electromagnetic wave interacts with $Fe_3O_4$ first when passing through (CNT/$Fe_3O_4$)/EP, part of the electromagnetic wave is absorbed due to hysteresis loss and natural resonance, and the rest will reach the CFs surface. Here, a part of electromagnetic wave is reflected back to (CNT/$Fe_3O_4$)/EP due to impedance mismatch, and the remaining part will pass through CFs to (CNT/$Fe_3O_4$)/EP on the other side [55]. CF@(CNT/$Fe_3O_4$)/EP with sandwich structure attenuates electromagnetic wave by multiple absorption, reflection and scattering processes and improves its EMI shielding performance [65–67]. Due to multilayer CF@(CNT/$Fe_3O_4$)/EP-CFs composite having more than one layer of CF@(CNT/$Fe_3O_4$)/EP, it provides more opportunities for electromagnetic wave propagation, so the above attenuation process of absorption–reflection/penetration–reabsorption for electromagnetic wave will be repeated many times and strengthens EMI shielding performance. In this process, since absorption is the main attenuation mode of electromagnetic wave, so the value of $SE_A$ is greater than $SE_R$. In addition, as the electromagnetic wave is absorbed, the heat (converted by the electromagnetic wave) generated by the electrical loss and magnetic loss accumulates inside the composite, which will cause the composite temperature increase.

Figure 6f shows the average values ($SE_{ave}$) of $SE_T$, $SE_R$ and $SE_A$ in X-band (8.2–12.4 GHz). When the amount of CNT/$Fe_3O_4$ is 0 (CF-0), the $SE_{ave}$ of $SE_T$, $SE_R$ and $SE_A$ are 10.56 dB, 2.03 dB and 8.53 dB, respectively. When the addition of CNT/$Fe_3O_4$ is increased to 0.64 mg/cm$^3$ (CF-8), compared with CF-0, the $SE_{ave}$ of $SE_T$, $SE_R$ and $SE_A$ are increased by 172%, 159% and 175% respectively. The reason is that CF-8 contains $Fe_3O_4$ while CF-0 does not, so the magnetic loss and dielectric loss for $Fe_3O_4$ are missing, which greatly reduces the electromagnetic wave absorption effect and leads to relatively poor EMI shielding performance [68,69].

Table 2 summarizes the λ, σ and EMI SE for some related polymer composites; it is observed that multilayer CF@(CNT/$Fe_3O_4$)/EP-CFs composite prepared by this work has good performances.

**Table 2.** Comparison of the λ, σ and EMI SE for some related composites.

| Sample | Loading | λ (W/(m·K)) | σ (S/cm) | Specific EMI SE (dB) | Thickness of Sample (mm) | Frequency Range (GHz) | Ref. |
|---|---|---|---|---|---|---|---|
| $Fe_3O_4$/CFs/Cement | 0.4 wt% CF + 5wt% $Fe_3O_4$ | - | - | 29.8 | 7 | 8.2–12.4 | [70] |
| CF@$Fe_3O_4$/EP | 20 wt% CF@$Fe_3O_4$ | - | - | 22.7 | 2 | 8.2–12.4 | [71] |
| PANI@nano-$Fe_3O_4$@CFs | 5 wt% of absorbing segments | - | - | 29 | 3 | 8.2–18 | [72] |
| Gt-MWCNT/SiC/HDPE | 23.1 vol% Gt-MWCNT + 11.3 vol% SiC | - | - | 14 | 2 | 8–12 | [73] |
| RGO@GF/EP | 40 wt% RGO-GF | - | - | 21.3 | 10 | 8.2–12.4 | [74] |
| PVDF@MWCNT/BN | 5 wt% MWCNT + 40 wt% BN | - | - | 4.34 | 2 | 8–12 | [65] |
| Ni@MWCNTs/HDPE | 3 wt% Ni@MWCNTs | - | - | 12 | 3 | 0.5–1.5 | [75] |
| CNTs-CFs/PF | 25 wt% CNTs and 40 wt% PF resin | 0.636 | - | - | - | - | [76] |
| Dry fabric/CNT mat | 1.06 wt% CNT | 1.386 | - | - | - | - | [17] |
| CF + CNT | 60 wt% CF and 40 wt% resin | - | 1.4 | - | - | - | [45] |
| CF@(CNT/$Fe_3O_4$)/EP-CFs | CF speed 0.01 m/s 0.045 wt% CNT/$Fe_3O_4$ | 1.1 | 7.7 | 30.5 | 2 | 8.2–12.4 | This work |

## 4. Conclusions

In this work, the multilayer CF@(CNT/$Fe_3O_4$)/EP-CFs composite is obtained by free arc dispersion-CFs surface spraying-rolling process method and structural design. Under circumstance of the content for CNT/$Fe_3O_4$ is very small, the (CNT/$Fe_3O_4$)/EP and CFs core sheath achieve thermal and electrical anisotropy and directional enhancement for multilayer CF@(CNT/$Fe_3O_4$)/EP-CFs composite, multilayer sandwich structure makes the EMI shielding performance better strengthened by multiple absorption–reflection/penetration–reabsorption of electromagnetic wave. From CF-0 to CF-8, the content of CNT/$Fe_3O_4$ only increases by 0.045 wt%, $\lambda_{\parallel}$ increases from 0.59 W/(m·K) to 1.1 W/(m·K), the growth rate is 86%, $\lambda_{\perp}$ only increases by 0.05 W/(m·K), and the maximum $\lambda_{\parallel}/\lambda_{\perp}$ is 2.9, $\sigma_{\parallel}$ increases from 6.2 S/cm to 7.7 S/cm, growth rate is 24%, $\sigma_{\perp}$ only increases by $0.7 \times 10^{-4}$ S/cm and EMI $SE_T$ increases by 196%, from 10.3 dB to 30.5 dB. This provides a new idea for enhancing CFs composite properties.

**Author Contributions:** Conceptualization, C.Z. and L.B.; methodology, C.Z.; software, S.S. and L.B.; validation, H.W. and D.Z.; formal analysis, S.S.; investigation, S.S.; data curation, L.B.; writing—original draft preparation, C.Z.; writing—review and editing, C.Z.; visualization, C.Z.; project administration, Y.H. and W.L.; funding acquisition, Y.H. All authors have read and agreed to the published version of the manuscript.

**Funding:** This research was funded by National Natural Science Foundation of China, grant number 52176076 and 51676103, Taishan Scholar Project of Shandong Province, grant number NO. ts20190937.

**Institutional Review Board Statement:** Not applicable.

**Informed Consent Statement:** Not applicable.

**Data Availability Statement:** Not applicable.

**Acknowledgments:** Authors acknowledge the support from Qingdao University of Science and Technology. This research made use of the Ministry of Science and Technology of polymer processing technology and the related thermal physics international cooperation base.

**Conflicts of Interest:** The authors declare no conflict of interest.

## References

1. Qiu, L.; Zheng, X.H.; Zhu, J.; Su, G.P.; Tang, D.W. The effect of grain size on the lattice thermal conductivity of an individual polyacrylonitrile-based carbon fiber. *Carbon* **2013**, *51*, 265–273. [CrossRef]
2. Wang, Q.F.; Ma, Y.; Liang, X.; Zhang, D.H.; Miao, M.H. Flexible supercapacitors based on carbon nanotube-$MnO_2$ nanocomposite film electrode. *Chem. Eng. J.* **2019**, *371*, 145–153. [CrossRef]
3. Cui, Y.; Qin, Z.H.; Wu, H.; Li, M.; Hu, Y.J. Flexible thermal interface based on self-assembled boron arsenide for high-performance thermal management. *Nat. Commun.* **2021**, *12*, 1284. [CrossRef]
4. Saylor, R.A.; Hersey, M.; West, A.; Buchanan, A.M.; Berger, S.N.; Nijhout, H.F.; Reed, M.C.; Best, J.; Hashemi, P. In vivo Hippocampal Serotonin Dynamics in Male and Female Mice: Determining Effects of Acute Escitalopram Using Fast Scan Cyclic Voltammetry. *Front. Neurosci.* **2019**, *13*, 103389. [CrossRef] [PubMed]
5. Shalaby, H.A.; Hassan, M.M.; Safar, S.S. Parametric study of shear strength of CFRP strengthened end-web panels. *Steel Compos. Struct.* **2019**, *31*, 159–172.
6. Kim, J.T.; Park, C.W.; Kim, B.J. A study on synergetic EMI shielding behaviors of Ni-Co alloy-coated carbon fibers-reinforced composites. *Synthetic. Met.* **2017**, *223*, 212–217. [CrossRef]
7. Xiao, S.H.; Zhou, X.; Deng, H.; Fu, Q. Preparation of elastic conductor with high stretchability and stable conductivity under strain via pre-stretching and spraying approach. *Compos. Commun.* **2021**, *24*, 100641. [CrossRef]
8. Ruan, K.P.; Zhong, X.; Shi, X.T.; Dang, J.J.; Gu, J.W. Liquid crystal epoxy resins with high intrinsic thermal conductivities and their composites: A mini-review. *Mater. Today Phys.* **2021**, *20*, 100456. [CrossRef]
9. Ruan, K.P.; Gu, J.W. Ordered Alignment of Liquid Crystalline Graphene Fluoride for Significantly Enhancing Thermal Conductivities of Liquid Crystalline Polyimide Composite Films. *Macromolecules* **2022**, *55*, 4134–4145. [CrossRef]
10. Song, P.; Liu, B.; Qiu, H.; Shi, X.T.; Cao, D.P.; Gu, J.W. MXenes for polymer matrix electromagnetic interference shielding composites: A review. *Compos. Commun.* **2021**, *24*, 100653. [CrossRef]
11. Qian, G.; Wu, B.; Qin, Z.; Li, X.; Zheng, Z.; Xia, R.; Qian, J. Enhanced Thermal Conductivity via In Situ Constructed CNT Aerogel Structure in Composites. *Adv. Mater. Interfaces* **2022**, *9*, 2102098. [CrossRef]
12. Liu, Z.F.; Chen, Z.H.; Yu, F. Enhanced thermal conductivity of microencapsulated phase change materials based on graphene oxide and carbon nanotube hybrid filler. *Sol. Energy Mater. Sol. Cells* **2019**, *192*, 72–80. [CrossRef]
13. Park, J.M.; Kwon, D.J.; Wang, Z.J.; Roh, J.U.; Lee, W.I.; Park, J.K.; DeVries, K.L. Effects of carbon nanotubes and carbon fiber reinforcements on thermal conductivity and ablation properties of carbon/phenolic composites. *Compos. Part. B-Eng.* **2014**, *67*, 22–29. [CrossRef]
14. Han, Z.D.; Fina, A. Thermal conductivity of carbon nanotubes and their polymer nanocomposites: A review. *Prog. Polym. Sci.* **2011**, *36*, 914–944. [CrossRef]
15. Jouni, M.; Djurado, D.; Massardier, V.; Boiteux, G. A representative and comprehensive review of the electrical and thermal properties of polymer composites with carbon nanotube and other nanoparticle fillers. *Polym. Int.* **2017**, *66*, 1237–1251. [CrossRef]
16. Deng, S.L.; Lin, Z.D.; Xu, B.F.; Lin, H.B.; Du, C.M. Effects of Carbon Fillers on Crystallization Properties and Thermal Conductivity of Poly(phenylene sulfide). *Polym.-Plast. Technol.* **2015**, *54*, 1017–1024. [CrossRef]
17. Shin, Y.C.; Novin, E.; Kim, H. Electrical and Thermal Conductivities of Carbon Fiber Composites with High Concentrations of Carbon Nanotubes. *Int. J. Precis. Eng. Man.* **2015**, *16*, 465–470. [CrossRef]
18. Kornev, K.G.; Halverson, D.; Korneva, G.; Gogotsi, Y.; Friedman, G. Magnetostatic interactions between carbon nanotubes filled with magnetic nanoparticles. *Appl. Phys. Lett.* **2008**, *92*, 233117. [CrossRef]
19. Korneva, G.; Ye, H.; Gogotsi, Y.; Halverson, D.; Friedman, G.; Bradley, J.C.; Kornev, K.G. Carbon Nanotubes Loaded with Magnetic Particles. *Nano Lett.* **2005**, *5*, 879–884. [CrossRef]
20. Kong, L.; Yin, X.W.; Yuan, X.Y.; Zhang, Y.J.; Liu, X.M.; Cheng, L.F.; Zhang, L.T. Electromagnetic wave absorption properties of graphene modified with carbon nanotube/poly(dimethyl siloxane) composites. *Carbon* **2014**, *73*, 185–193. [CrossRef]
21. Chu, Z.Y.; Cheng, H.F.; Xie, W.; Sun, L.K. Effects of diameter and hollow structure on the microwave absorption properties of short carbon fibers. *Ceram. Int.* **2012**, *38*, 4867–4873. [CrossRef]
22. Liu, Y.; Zhang, Z.Q.; Xiao, S.T.; Qiang, C.W.; Tian, L.L.; Xu, J.C. Preparation and properties of cobalt oxides coated carbon fibers as microwave-absorbing materials. *Appl. Surf. Sci.* **2011**, *257*, 7678–7683. [CrossRef]
23. Qiang, C.W.; Xu, J.C.; Zhang, Z.Q.; Tian, L.L.; Xiao, S.T.; Liu, Y.; Xu, P. Magnetic properties and microwave absorption properties of carbon fibers coated by $Fe_3O_4$ nanoparticles. *J. Alloys Compd.* **2010**, *506*, 93–97. [CrossRef]
24. Wu, X.Y.; Tu, T.X.; Dai, Y.; Tang, P.P.; Zhang, Y.; Deng, Z.M.; Li, L.L.; Zhang, H.B.; Yu, Z.Z. Direct Ink Writing of Highly Conductive MXene Frames for Tunable Electromagnetic Interference Shielding and Electromagnetic Wave-Induced Thermochromism. *Nano-Micro. Lett.* **2021**, *13*, 148. [CrossRef] [PubMed]
25. Deng, Z.M.; Tang, P.P.; Wu, X.Y.; Zhang, H.B.; Yu, Z.Z. Superelastic, Ultralight, and Conductive $Ti_3C_2T_x$ MXene/Acidified Carbon Nanotube Anisotropic Aerogels for Electromagnetic Interference Shielding. *ACS. Appl. Mater. Interfaces* **2021**, *13*, 20539–20547. [CrossRef]
26. Gao, J.S.; Wang, H.H.; Zhou, Y.; Liu, Z.M.; He, Y. Self-template and in-situ synthesis strategy to construct $MnO_2/Mn_3O_4$@Ni-Co/GC nanocubes for efficient microwave absorption properties. *J. Alloy Compd.* **2022**, *892*, 162151. [CrossRef]

27. Liang, Y.Y.; Li, Y.G.; Wang, H.L.; Zhou, J.G.; Wang, J.; Regier, T.; Dai, H.J. $Co_3O_4$ nanocrystals on graphene as a synergistic catalyst for oxygen reduction reaction. *Nat. Mater.* **2011**, *10*, 780–786. [CrossRef]
28. Eerenstein, W.; Mathur, N.D.; Scott, J.F. Multiferroic and magnetoelectric materials. *Nature* **2006**, *442*, 759–765. [CrossRef]
29. Ohlan, A.; Singh, K.; Chandra, A.; Dhawan, S.K. Microwave absorption properties of conducting polymer composite with barium ferrite nanoparticles in 12.4–18 GHz. *Appl. Phys. Lett.* **2008**, *93*, 053114. [CrossRef]
30. Xu, H.F.; Zhang, H.J.; Lv, T.; Wei, H.W.; Song, F. Study on $Fe_3O_4$/polyaniline electromagnetic composite hollow spheres prepared against sulfonated polystyrene colloid template. *Colloid. Polym. Sci.* **2013**, *291*, 1713–1720. [CrossRef]
31. Kumar, A.; Singh, A.P.; Kumari, S.; Srivastava, A.K.; Bathula, S.; Dhawan, S.K.; Duttab, P.K.; Dhar, A. EM shielding effectiveness of Pd-CNT-Cu nanocomposite buckypaper. *J. Mater. Chem. A* **2015**, *3*, 13986–13993. [CrossRef]
32. Li, D.X.; Zhou, X.W.; Guo, X.J.; Yuan, B.; Liu, Y.J.; Ortega, C.M.; Sun, L.; Liu, Z. A novel $Fe_3O_4$/buckypaper composite as free-standing anode for lithium-ion battery. *J. Alloy Compd.* **2016**, *657*, 109–114. [CrossRef]
33. Chaudhary, A.; Kumar, R.; Teotia, S.; Dhawan, S.K.; Dhakate, S.R.; Kumari, S. Integration of MCMBs/MWCNTs with $Fe_3O_4$ in a flexible and light weight composite paper for promising EMI shielding applications. *J. Mater. Chem. C* **2017**, *5*, 322–332. [CrossRef]
34. Rao, B.V.B.; Chengappa, M.; Kale, S.N. Lightweight, flexible and thin $Fe_3O_4$-loaded, functionalized multi walled carbon nanotube buckypapers for enhanced X-band electromagnetic interference shielding. *Mater. Res. Express.* **2017**, *4*, 045012.
35. Li, S.L.; He, Y.; Jing, C.W.; Gong, X.B.; Cui, L.L.; Cheng, Z.Y.; Zhang, C.Q.; Nan, F. A novel preparation and formation mechanism of carbon nanotubes aerogel. *Carbon Lett.* **2018**, *28*, 16–23.
36. Li, S.L.; Zhang, C.Q.; He, Y.; Feng, M.; Ma, C.; Cui, Y. Multi-interpolation mixing effects under the action of micro-scale free arc. *J. Mater. Process. Technol.* **2019**, *271*, 645–650. [CrossRef]
37. Uetani, K.; Takahashi, K.; Watanabe, R.; Tsuneyasu, S.; Satoh, T. Thermal Diffusion Films with In-Plane Anisotropy by Aligning Carbon Fibers in a Cellulose Nanofiber Matrix. *ACS. Appl. Mater. Inter.* **2022**, *14*, 33903–33911. [CrossRef]
38. Nagai, H.; Fujita, K.; Urabe, K.; Iwashita, N. FEM analysis of flexural modulus of carbon fiber monofilament considering anisotropy. *Adv. Compos. Mater.* **2022**, *31*, 137–150. [CrossRef]
39. Danlee, Y.; Bailly, C.; Huynen, I. Thin and flexible multilayer polymer composite structures for effective control of microwave electromagnetic absorption. *Compos. Sci. Technol.* **2014**, *100*, 182–188. [CrossRef]
40. Yang, L.; Fan, H.L.; Liu, J.; Ma, Y.; Zheng, Q. Hybrid lattice-core sandwich composites designed for microwave absorption. *Mater. Des.* **2013**, *50*, 863–871. [CrossRef]
41. Agarwal, P.R.; Kumar, R.; Kumari, S.; Dhakate, S.R. Three-dimensional and highly ordered porous carbon-$MnO_2$ composite foam for excellent electromagnetic interference shielding efficiency. *RSC Adv.* **2016**, *6*, 100713–100722. [CrossRef]
42. Kumar, R.; Dhakate, S.R.; Saini, P.; Mathur, R.B. Improved electromagnetic interference shielding effectiveness of light weight carbon foam by ferrocene accumulation. *RSC Adv.* **2013**, *3*, 4145–4151. [CrossRef]
43. Zhang, Y.L.; Ruan, K.P.; Gu, J.W. Flexible Sandwich-Structured Electromagnetic Interference Shielding Nanocomposite Films with Excellent Thermal Conductivities. *Small* **2021**, *17*, 10. [CrossRef] [PubMed]
44. Al-Saleh, M.H.; Saadeh, W.H.; Sundararaj, U. EMI shielding effectiveness of carbon based nanostructured polymeric materials: A comparative study. *Carbon* **2013**, *60*, 146–156. [CrossRef]
45. Li, S.L.; Zhang, C.Q.; Fu, J.F.; Zhou, Y.S.; Sun, J.Q.; He, Y.; Nan, F.; Yu, Z.Z. Interfacial modification of carbon fiber by carbon nanotube gas-phase dispersion. *Compos. Sci. Technol.* **2020**, *195*, 108196. [CrossRef]
46. Zhang, M.; Ding, L.; Zheng, J.; Liu, L.B.; Alsulami, H.; Kutbi, M.A.; Xu, J.L. Surface modification of carbon fibers with hydrophilic $Fe_3O_4$ nanoparticles for nickel-based multifunctional composites. *Appl. Surf. Sci.* **2020**, *509*, 145348. [CrossRef]
47. Panupakorn, P.; Chaichana, E.; Praserthdam, P.; Jongsomjit, B. Polyethylene/Clay Nanocomposites Produced by In Situ Polymerization with Zirconocene/MAO Catalyst. *J. Nanomater.* **2013**, *2013*, 154874. [CrossRef]
48. Mueller, F.; Bresser, D.; Paillard, E.; Winter, M.; Passerini, S. Influence of the carbonaceous conductive network on the electrochemical performance of $ZnFe_2O_4$ nanoparticles. *J. Power Sources.* **2013**, *236*, 87–94. [CrossRef]
49. Pan, Y.; Zeng, W.J.; Li, L.; Zhang, Y.Z.; Dong, Y.N.; Ye, K.; Cheng, K.; Cao, D.X.; Wang, G.L.; Lucht, B.L. Surfactant assisted, one-step synthesis of $Fe_3O_4$ nanospheres and further modified $Fe_3O_4$/C with excellent lithium storage performance. *J. Electroanal. Chem.* **2018**, *810*, 248–254. [CrossRef]
50. Zhang, J.; Tan, P.H.; Zhao, W.J.; Lu, J.; Zhao, J.H. Raman study of ultrathin $Fe_3O_4$ films on GaAs(001) substrate: Stoichiometry, epitaxial orientation and strain. *J. Raman Spectrosc.* **2011**, *42*, 1388–1391. [CrossRef]
51. Wang, L.P.; Huang, Y.B.; Lai, Y.H. Surface enhanced Raman scattering activity of dual-functional $Fe_3O_4$/Au composites. *Appl. Surf. Sci.* **2018**, *435*, 290–296. [CrossRef]
52. Belin, T.; Epron, F. Characterization methods of carbon nanotubes: A review. *Mat. Sci. Eng. B* **2005**, *119*, 105–118. [CrossRef]
53. Pan, L.; Tang, J.; Chen, Y.H. Synthesis of $Fe_3O_4$, $Fe_2O_3$, Ag/$Fe_3O_4$ and Ag/$Fe_2O_3$ nanoparticles and their electrocatalytic properties. *Sci. China Chem.* **2013**, *56*, 362–369. [CrossRef]
54. Pop, E.; Mann, D.; Wang, Q.; Goodson, K.; Da, H. Thermal Conductance of an Individual Single-Wall Carbon Nanotube above Room Temperature. *Nano Lett.* **2006**, *6*, 96–100. [CrossRef]
55. Guo, Y.Q.; Qiu, H.; Ruan, K.P.; Wang, S.S.; Zhang, Y.L.; Gu, J.W. Flexible and insulating silicone rubber composites with sandwich structure for thermal management and electromagnetic interference shielding. *Compos. Sci. Technol.* **2022**, *219*, 109253. [CrossRef]
56. Bily, M.A.; Kwon, Y.W.; Pollak, R.D. Study of Composite Interface Fracture and Crack Growth Monitoring Using Carbon Nanotubes. *Appl. Compos. Mater.* **2010**, *17*, 347–362. [CrossRef]

57. Zheng, Y.D.; Wang, R.; Dong, X.Y.; Wu, L.X.; Zhang, X. High Strength Conductive Polyamide 6 Nanocomposites Reinforced by Prebuilt Three-Dimensional Carbon Nanotube Networks. *ACS Appl. Mater. Interfaces* **2018**, *10*, 28103–28111. [CrossRef]
58. Kim, P.; Shi, L.; Majumdar, A.; Mceuen, P.L. Thermal Transport Measurements of Individual Multiwalled NanoTubes. *Phys. Rev. Lett.* **2001**, *87*, 215502. [CrossRef]
59. Kausar, A.; Ahmad, S.; Salman, S.M. Effectiveness of Polystyrene/Carbon Nanotube Composite in Electromagnetic Interference Shielding Materials: A Review. *Polym.-Plast. Technol.* **2017**, *56*, 1027–1042. [CrossRef]
60. Ebbesen, T.W.; Lezec, H.J.; Hiura, H.; Bennett, J.W.; Ghaemi, H.F.; Thio, T. Electrical conductivity of individual carbon nanotubes. *Nature* **1996**, *382*, 54–56. [CrossRef]
61. Haghgoo, M.; Ansari, R.; Hassanzadeh-Aghdam, M.K. Prediction of electrical conductivity of carbon fiber-carbon nanotube-reinforced polymer hybrid composites. *Compos. Part B-Eng.* **2019**, *167*, 728–735. [CrossRef]
62. Ji, X.Y.; Matsuo, S.; Sottos, N.R.; Cahill, D.G. Anisotropic thermal and electrical conductivities of individual polyacrylonitrile-based carbon fibers. *Carbon* **2022**, *197*, 1–9. [CrossRef]
63. Bagheli, S.; Fadafan, H.K.; Orimi, R.L.; Ghaemi, M. Synthesis and experimental investigation of the electrical conductivity of water based magnetite nanofluids. *Powder Technol.* **2015**, *274*, 426–430. [CrossRef]
64. Qiao, M.T.; Li, J.X.; Wei, D.; He, X.W.; Lei, X.F.; Wei, J.; Zhang, Q.Y. Chain-like $Fe_3O_4$@void@m$SiO_2$@$MnO_2$ composites with multiple porous shells toward highly effective microwave absorption application. *Micropor. Mesopor. Mat.* **2021**, *314*, 110867. [CrossRef]
65. Zhang, P.; Ding, X.; Wang, Y.Y.; Gong, Y.; Zheng, K.; Chen, L.; Tian, X.Y.; Zhang, X. Segregated double network enabled effective electromagnetic shielding composites with extraordinary electrical insulation and thermal conductivity. *Compos. Part A-Appl. S.* **2019**, *117*, 56–64. [CrossRef]
66. Feng, C.P.; Wan, S.S.; Wu, W.C.; Bai, L.; Bao, R.Y.; Liu, Z.Y.; Yang, M.B.; Chen, J.; Yang, W. Electrically insulating, layer structured SiR/GNPs/BN thermal management materials with enhanced thermal conductivity and breakdown voltage. *Compos. Sci. Technol.* **2018**, *167*, 456–462. [CrossRef]
67. Huang, S.; Wang, L.; Li, Y.C.; Liang, C.B.; Zhang, J.L. Novel $Ti_3C_2T_x$ MXene/epoxy intumescent fire-retardant coatings for ancient wooden architectures. *J. Appl. Polym. Sci.* **2021**, *138*, 50649. [CrossRef]
68. Liu, Z.S.; Zhang, Y.; Zhang, H.B.; Dai, Y.; Liu, J.; Li, X.F.; Yu, Z.Z. Electrically conductive aluminum ion-reinforced MXene films for efficient electromagnetic interference shielding. *J. Mater. Chem. C* **2020**, *8*, 1673–1678. [CrossRef]
69. Luo, J.Q.; Zhao, S.; Zhang, H.B.; Deng, Z.M.; Li, L.L.; Yu, Z.Z. Flexible, stretchable and electrically conductive MXene/natural rubber nanocomposite films for efficient electromagnetic interference shielding. *Compos. Sci. Technol.* **2019**, *182*, 107754. [CrossRef]
70. Liu, Z.F.; Ge, H.Y.; Wu, J.M.; Chen, J. Enhanced electromagnetic interference shielding of carbon fiber/cement composites by adding ferroferric oxide nanoparticles. *Constr. Build. Mater.* **2017**, *151*, 575–581. [CrossRef]
71. Gholampour, M.; Movassagh-Alanagh, F.; Salimkhani, H. Fabrication of nano-$Fe_3O_4$ 3D structure on carbon fibers as a microwave absorber and EMI shielding composite by modified EPD method. *Solid State Sci.* **2017**, *64*, 51–61. [CrossRef]
72. Movassagh-Alanagh, F.; Bordbar-Khiabani, A.; Ahangari-Asl, A. Three-phase PANI@nano-$Fe_3O_4$@CFs heterostructure: Fabrication, characterization and investigation of microwave absorption and EMI shielding of PANI@nano-$Fe_3O_4$@CFs/epoxy hybrid composite. *Compos. Sci. Technol.* **2017**, *150*, 65–78. [CrossRef]
73. Zhang, X.M.; Zhang, J.J.; Xia, L.C.; Wang, J.F.; Li, C.H.; Xu, F.; Zhang, X.L.; Wu, H.; Guo, S.Y. Achieving high-efficiency and robust 3D thermally conductive while electrically insulating hybrid filler network with high orientation and ordered distribution. *Chem. Eng. J.* **2018**, *334*, 247–256. [CrossRef]
74. Wan, X.R.; Lu, H.; Kang, J.F.; Li, S.; Yue, Y.L. Preparation of graphene-glass fiber-resin composites and its electromagnetic shielding performance. *Compos. Interface* **2018**, *25*, 883–900. [CrossRef]
75. Yim, Y.J.; Rhee, K.Y.; Park, S.J. Electromagnetic interference shielding effectiveness of nickel-plated MWCNTs/high-density polyethylene composites. *Compos. Part B-Eng.* **2016**, *98*, 120–125. [CrossRef]
76. Feng, A.; Jia, Z.; Yu, Q.; Zhang, H.; Wu, G. Preparation and Characterization of Carbon Nanotubes/Carbon Fiber/Phenolic Composites on Mechanical and Thermal Conductivity Properties. *Nano* **2018**, *13*, 1850037. [CrossRef]

*materials*

*Article*

# Failure Analysis of Abnormal Cracking of the Track Circuit Reader Antenna Baffle for High-Speed Trains

**Chang Su, Tong-Tong Bi and Zhen-Guo Yang ***

Department of Materials Science, Fudan University, Shanghai 200433, China; 19110300018@fudan.edu.cn (C.S.); 20110300038@fudan.edu.cn (T.-T.B.)
* Correspondence: zgyang@fudan.edu.cn; Tel.: +86-021-31243658

**Abstract:** The track circuit reader (TCR) is an important part of train control systems. This paper reports a failure of the TCR antenna baffle, which is used to prevent the TCR antenna from being struck by foreign objects. The designed service life of the baffle is 4.8 million kilometers, but serious cracking was found during routine maintenance after only 0.67 million kilometers of operation. In order to avoid the hidden danger brought by the incident to the safe operation of the train, it is necessary to conduct a complete failure analysis of the failed TCR antenna baffle. Therefore, a comprehensive investigation of the base material, cleaning agents, crack morphologies, etc., was carried out, and the failure environment of the antenna baffle was verified by experiment. The final results show that the environmental stress cracking is the root cause of the failed antenna baffle, and the multiple bubbles produced by the formed process of the antenna baffle are another important cause. According to the conclusions, the solutions to prevent the reoccurrence of such failures are proposed. After these solutions are adopted, the number of failed antenna baffles is greatly reduced, which fully proves the correctness of this analysis.

**Keywords:** polyurethane rubber; antenna baffle; environmental stress cracking; bubble; failure analysis

**Citation:** Su, C.; Bi, T.-T.; Yang, Z.-G. Failure Analysis of Abnormal Cracking of the Track Circuit Reader Antenna Baffle for High-Speed Trains. *Materials* **2023**, *16*, 722. https://doi.org/10.3390/ma16020722

Academic Editor: Juraj Gerlici

Received: 12 December 2022
Revised: 6 January 2023
Accepted: 10 January 2023
Published: 11 January 2023



## 1. Introduction

The track circuit reader (TCR) is a signal subsystem used on some electric multiple unit (EMU) trains with speeds over 300 km/h. As an important part of the train control system, it reads the relevant track circuit information code and provides normal or braking information for the onboard safety computer. According to the output information of the TCR and the information of the ground transponder, the train control system ensures the security of train operation. The TCR antenna is the exposed part of the TCR device. In order to prevent the TCR antenna from being struck by foreign objects when the train is running, a baffle is installed on the side of the TCR antenna to protect it. Due to excellent physical and chemical properties, polyurethane rubber (PUR) is often used as TCR antenna baffles (hereinafter referred to as antenna baffle). Once the baffle fails, the TCR antenna will be exposed to a dangerous environment, which will bring danger to not only the normal operation of the TCR device, but also the safe operation of the train. In addition, if the antenna baffle cracks and falls on the track, it will bring derailment risk to high-speed trains. Therefore, the failure of the antenna baffle deserves attention.

Due to its high hardness, good strength, high elasticity, wear resistance, tear resistance, and aging resistance [1,2], PUR is widely used [3–6]. However, it turns out that like other polymer materials [7], PUR also has its drawbacks. Under some specific circumstances, these drawbacks are magnified. Ren et al. [8] studied the causes of inclusion defects on the surface of strip steel caused by PU rolls. In order to clarify the source of inclusions, morphology, and chemical composition of inclusions, wear debris and adhesion substances on PU rollers were analyzed in detail. Finally, it was found that the wear of the PU roller is the root cause leading to the inclusions. Junik et al. [9] researched the impact of the hardness on the selected mechanical properties of rigid PU elastomers commonly used in suspension systems. The results show that the hardness plays an important role in the fracture sensitivity of PUR material, and in a certain hardness range, PUR with higher hardness has higher toughness, wear resistance, and fatigue resistance. Wen et al. [10] found that electrostatic cumulative discharges on the surfaces of PU elastomers could lead to accidents and disasters, such as equipment failures, fires, and explosions, in industries. In order to solve this problem, they adopted the method of embedding copper foam in PU to form a copper foam-based PU composite, which greatly improved the electrostatic protection performance and wear resistance of PU. Cristiano et al. [11] developed a new methodology to study the failure of elastomers in a confined geometry and applied this new methodology to model end-linked PU elastomers. In situ experiments show that the failure of elastomers is due to the growth of a single cavity nucleated in the region of maximum hydrostatic stress. Comparison between different elastomers shows that the material containing both entanglements and crosslinks is more resistant to cavitation relative to its elastic modulus. In addition to the factors mentioned above, PUR is also affected by fatigue cracking [12,13], thermal oxidation aging [14], physical swelling [15,16], and so on. These factors have also been found in the failure of other rubber materials [17–19]. These studies mentioned above only focus on a specific property of PUR, and only propose corresponding solutions from a certain aspect. However, most PURs fail in complex environments, and the failure causes are often multiple rather than single. Therefore, studying the failure of PUR in specific working conditions can not only solve the actual failure problem, but also provide a reference for similar failure cases in the future.

In this paper, the material and performance of the failed baffle are analyzed by various testing methods, and its fracture surface is carefully observed by macro- and micro-observation. In addition, the root cause of the failure was verified again by experiment. Finally, it was found that the antenna baffle was exposed to the 602 cleaning agent, which caused environmental cracking stress. This is the root cause of the failed baffle. Furthermore, it was found that there were some bubbles inside the baffle, which led to the decline of its mechanical properties and made it more prone to capillary action. This is an important cause of the failed baffle. According to the above reasons, reasonable solutions are proposed. After these solutions are applied to the actual protection of the baffle, the number of failed antenna baffles is greatly reduced. This study can not only solve the failure problem of TCR antenna baffles for high-speed trains, but also help to update the understanding of the failure mechanism of PUR.

## 2. Background

The antenna baffle is used on the front bogie of the high-speed train, and the specific location is shown in Figure 1a. In Figure 1a, it can also be seen that there is an obvious blue line on the train body, which divides the body into upper and lower parts. Figure 1b shows the appearance of the antenna baffle. For the convenience of description, the side facing the antenna is defined as the back side, and the other side is defined as the front side. Figure 1c shows the failure of the baffle, and it can be seen that there are obvious cracks in the fixed part of the baffle. Figure 2 is an installation diagram of the antenna baffle. It shows that the antenna baffle is fixed on the bracket by two stainless steel splints and five bolts with tightening torque of 10 Nm. It is worth mentioning that the antenna baffle only exists on one side of the TCR antenna. Figure 3 shows the appearance of the light cyan new baffle and the black-brown failed baffle, which are named Sample 1# and Sample 2# in turn for the convenience of subsequent description. During the maintenance, there were two cases of antenna baffle cracking. In order to eliminate this hidden danger, the company of the high-speed train immediately carried out a special census involving 282 trains and found 21 cases of similar problems. This shows that the abnormal cracking of the antenna baffle is universal and needs to be solved urgently. In addition, the baffle was designed to have a service life of 4.8 million kilometers, but it had actually only operated 0.67 million kilometers before serious cracking occurred. Obviously, this is a premature, abnormal failure, which will bring hidden dangers to the safe operation of the train. Therefore, it is very necessary to study the cause of the premature failure.

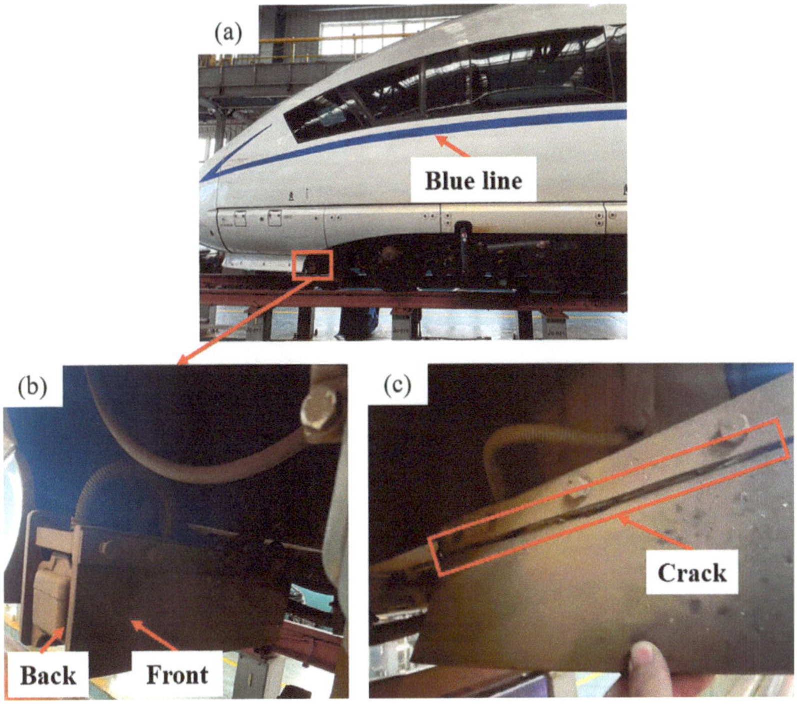


**Figure 1.** Specific location and appearance of the antenna baffle: (**a**) specific location of the baffle and the blue line; (**b**) front and back sides of the antenna baffle; (**c**) crack on the baffle.

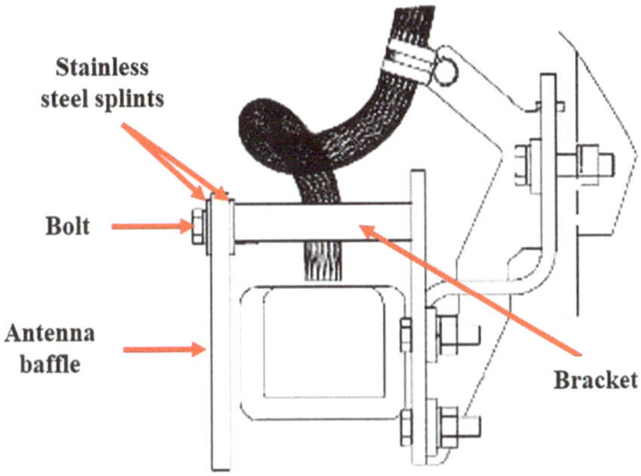


**Figure 2.** Installation diagram of the antenna baffle.

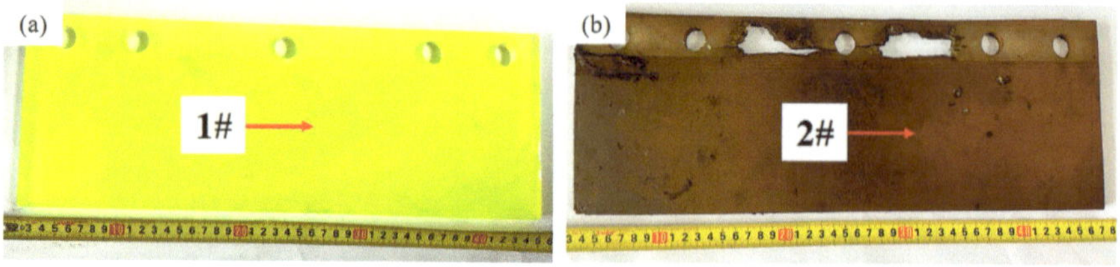


**Figure 3.** Appearance of the antenna baffles: (**a**) appearance of the new baffle; (**b**) appearance of the failed baffle.

## 3. Investigations and Discussions

### 3.1. Characterization Analysis

#### 3.1.1. Macroscopic Observation

Figure 4a shows the front appearance of Sample 2#. There is a lot of dust stuck to the surface of the baffle, and there is a raised strip of adhesive on the right side of the baffle, which should have been reinforced with adhesive tape after the crack was found. After partial enlargement of the front side, the cracking failure appearance is more obvious, as illustrated in Figure 4c. Figure 4b reveals the back appearance of Sample 2#. There is also adhesion on the left side. After partial enlargement of the back side, the cracking failure appearance is also obvious, as shown in Figure 4d.

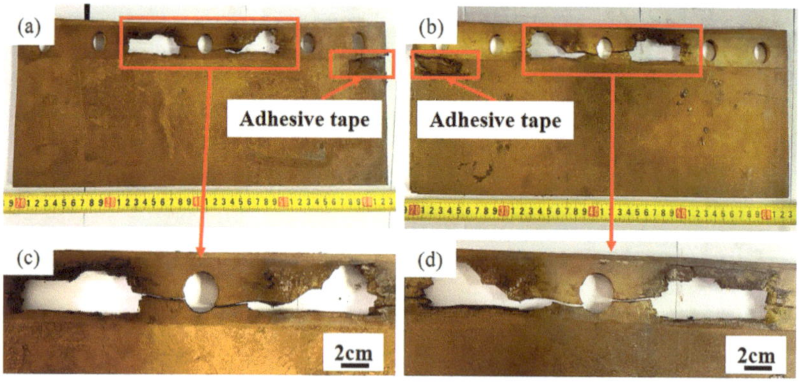


**Figure 4.** Appearance of Sample 2#: (**a**) overall appearance of the front side; (**b**) overall appearance of the back side; (**c**) cracks on the front side after partial enlargement; (**d**) cracks on the back side after partial enlargement.

#### 3.1.2. Three-Dimensional Stereomicroscope (3D-SM) Observation

Figure 5a reveals the front overall appearance of Sample 2#. For the convenience of observation, the sample in the red box is cut and sampled. The front appearance of the sample after sampling is shown in Figure 5b. After placing the sample vertically (Figure 5c), the cross-section observed under 3D-SM is shown in Figure 5d–f. The three sections are greasy and bright with a lot of dust, and no obvious crack initiation point is observed. In addition, when the cross-section of the sample is touched by hand, it is obviously sticky and has low hardness (the fingernail can easily pierce into the sample), which indicates that the material has obvious performance degradation.

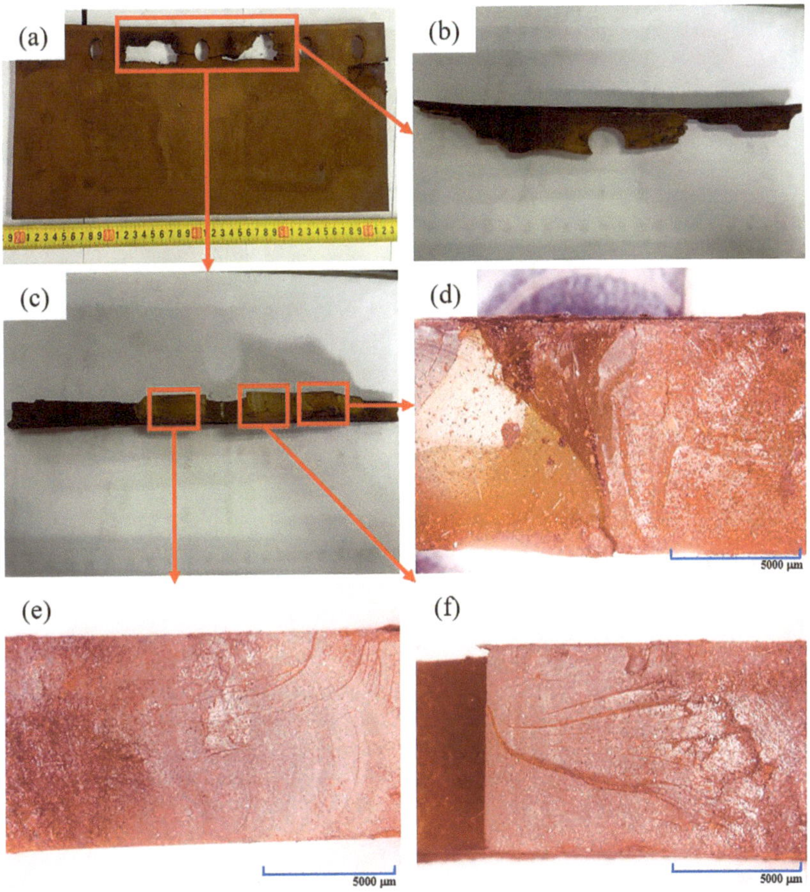


**Figure 5.** Cross-section appearance of Sample 2#: (**a**) overall appearance of Sample 2#; (**b**) sample obtained after sampling; (**c**) samples placed vertically; (**d**) appearance of the first section; (**e**) appearance of the second section; (**f**) appearance of the third section.

3.1.3. Scanning Electron Microscope (SEM) Observation

In order to facilitate the observation of its morphology under SEM, Sample 2# was sonicated in deionized water for 2 h, and then dried with a blower in cold air before observation.

Figure 6a is the fractograph of Sample 2#. Figure 6b–d shows that the fracture of Sample 2# is characterized by sticky, cracking, and adhesion of cracks. Figure 6c shows that the main crack has many branch cracks, which is a typical environmental stress cracking (ESC) morphology. This indicates that Sample 2# was seriously affected by the external environment, resulting in a significant decrease in its performance.

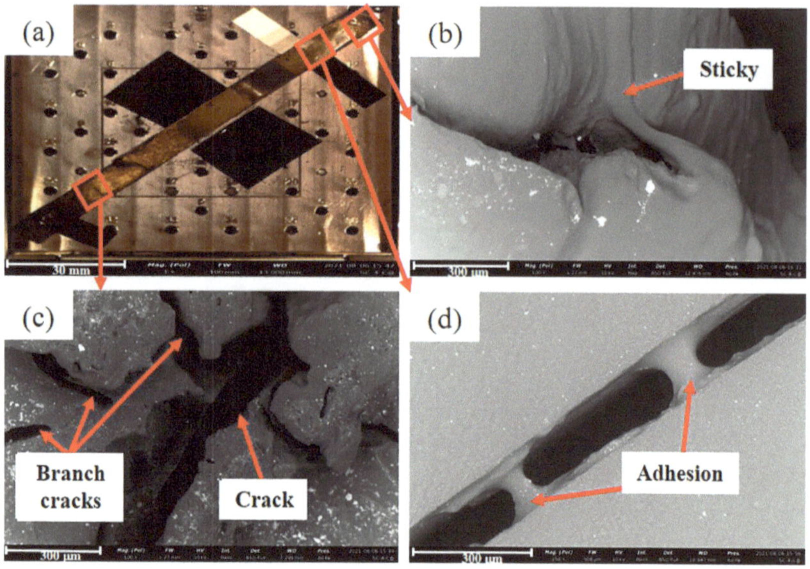


**Figure 6.** Fractograph of Sample 2#: (**a**) overall macro morphology of the section; (**b**) sticky; (**c**) main crack and branch cracks; (**d**) adhesion of the crack.

### 3.2. Material Analysis

#### 3.2.1. Attenuated Total Internal Reflectance Fourier Transform Infrared Spectroscopy (ATR-FTIR) Analysis of the Antenna Baffle

In order to determine whether new substances were produced before and after the failure, Samples 1# and 2# were analyzed by ATR-FTIR, and the analysis results are shown in Figure 7. Comparing the test results of the two samples, it can be found that the characteristic peaks of Sample 2# completely coincide with those of Sample 1#, which indicates that no chemical changes have taken place in Sample 2#. In addition, Figure 7 also shows that there are nine obvious characteristic peaks in both samples. The meaning of each characteristic peak is shown in Table 1 [20]. The existence of these characteristic peaks proves that the TCR antenna baffle is a synthetic polyurethane rubber [21,22].

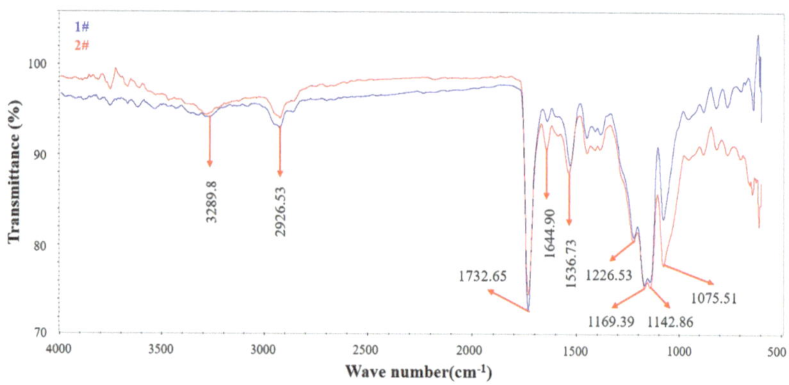


**Figure 7.** Infrared spectra of Samples 1# and 2#.

**Table 1.** Meanings of each characteristic peak [20–22].

| Serial Number | Peak | Meaning of the Peak |
|---|---|---|
| 1 | 3289 $cm^{-1}$ | Stretching vibration of N-H |
| 2 | 2926 $cm^{-1}$ | Antisymmetric stretching vibration of $-CH_2$ [23] |
| 3 | 1732 $cm^{-1}$ | Stretching vibration of ester carbonyl C=O |
| 4 | 1644 $cm^{-1}$ | Bending vibration peak of C=O in urea |
| 5 | 1536 $cm^{-1}$ | Band of amide II |
| 6 | 1226 $cm^{-1}$ | Stretching vibration of C-O |
| 7 | 1169 $cm^{-1}$ | Stretching vibration of C-OH |
| 8 | 1142 $cm^{-1}$ | Stretching vibration of C-OH |
| 9 | 1075 $cm^{-1}$ | Stretching vibration of C-O in the ether group |

3.2.2. Nuclear Magnetic Resonance (NMR) Analysis of the Antenna Baffle

The NMR test results of the antenna baffle (Table 2) show that compared with the new Sample 1#, the crosslinking density of the failed Sample 2# is greatly reduced and the dispersion coefficient of the failed Sample 2# is higher. This indicates that part of the crosslinking bonds in Sample 2# is broken, and the breakage of the crosslinking bonds is not uniform. Since the FTIR test results show that the antenna baffle has no obvious chemical changes, it can be reasonably speculated that the antenna baffle may be affected by environmental stress cracking (ESC), resulting in the breakage of its crosslinking bonds.

**Table 2.** Crosslinking density of Samples 1# and 2# ($10^{-4}$ mol/mL).

| Sample | First | Second | Third | Average | Dispersion Coefficients/% |
|---|---|---|---|---|---|
| 1# | 19.221 | 18.901 | 19.058 | 19.060 | 0.840 |
| 2# | 7.036 | 6.533 | 6.600 | 6.723 | 4.063 |

3.2.3. Thermogravimetric Analysis (TGA) of the Antenna Baffle

Figure 8 shows the thermogravimetric curve of Samples 1# and 2#. It shows that the final mass loss of Samples 1# and 2# is 89.79% (solid line in Figure 8), which indicates that there is no significant difference in mass loss between these two samples. However, by observing their mass loss rate curves (dotted line in Figure 8), it can be found that the temperature when the mass loss rate of Sample 1# reaches its maximum is 417.50 °C, and Sample 2# reaches its maximum is 365.33 °C. The difference is 52°C. That is to say, compared with Sample 1#, the temperature of Sample 2# when the mass loss rate reaches the maximum is earlier, which indicates that the heat resistance of Sample 2# is significantly reduced. Combined with the fact that the crosslinking density of Sample 2# is significantly lower than that of Sample 1#, it is believed that the ESC occurred after Sample 2# was eroded by solvent, which resulted in partial breakage of its crosslinking bonds, enhanced molecular fluidity, and decreased heat resistance.

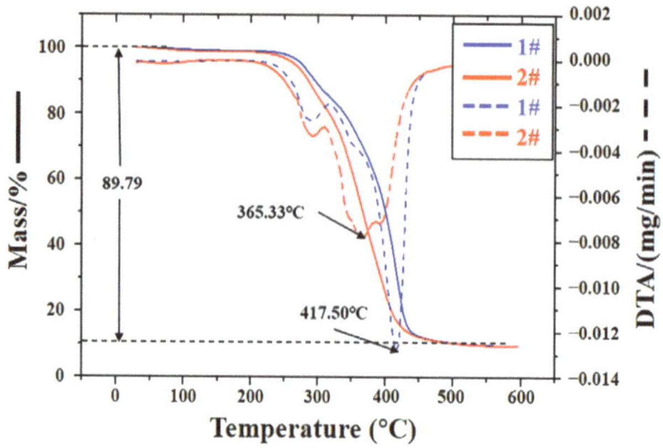


**Figure 8.** TGA curves of Samples 1# and 2#.

3.2.4. Differential Scanning Calorimetry (DSC) Analysis of the Antenna Baffle

In order to understand the change of the glass transition temperature (Tg) before and after the failure of the antenna baffle, a DSC test was carried out on the two samples respectively, and the test results are shown in Figure 9. The Tg of Samples 1# and 2# is −21.2 °C and −22.4 °C respectively, which indicates that the Tg of Sample 1# is higher than that of Sample 2#. Combined with the NMR and TGA test results, it can be reasonably speculated that Sample 2# was affected by ESC, resulting in the breakage of its crosslinking bonds and the easier movement of the molecular chain, thus leading to the decrease in its Tg.

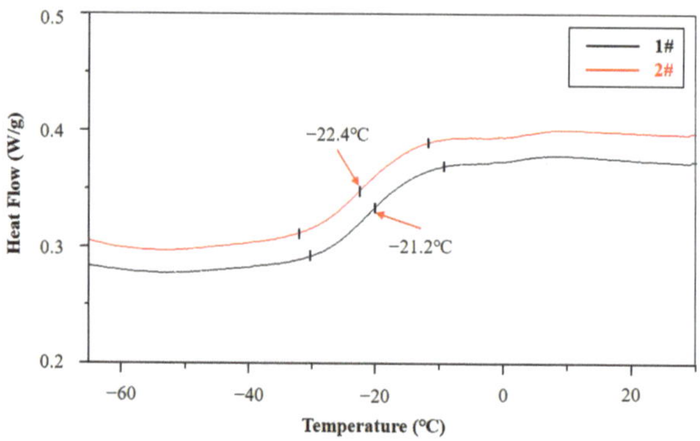


**Figure 9.** DSC curves of Samples 1# and 2#.

3.2.5. Hardness Analysis of the Antenna Baffle

The Shore A hardness tester was used to test the surface hardness of Samples 1# and 2#, and the test results are shown in Table 3.

**Table 3.** Hardness of Samples 1# and 2# (HA).

| Sample | First | Second | Third | Average |
|---|---|---|---|---|
| 1# | 83.5 | 84.0 | 82.5 | 83.3 |
| 2# | 67.5 | 68.0 | 68.5 | 68.0 |

Table 3 indicates that the hardness of Sample 2# is significantly lower than that of Sample 1#, indicating that the surface of Sample 2# is obviously softened. Combined with the test results of NMR, TGA, and DSC, it can be determined that Sample 2# is affected by ESC, resulting in a great decrease in its surface hardness.

### 3.2.6. Gas Chromatography–Mass Spectrometry (GC-MS) Analysis of the Cleaning Agents

The high-speed train is cleaned and maintained regularly, and four cleaning agents are used during cleaning. Three of them, which are named alkaline, neutral, and acidic cleaning agents, are used to clean the train body. Another cleaning agent named 602 is specially designed to clean the metal fixing parts of the bogie. It is worth mentioning that these cleaning agents are likely to contact the antenna baffle. To analyze the composition of these cleaning agents, GC-MS tests were carried out.

The analysis results of the alkaline cleaning agent are shown in Figure 10. It indicates that no matching organic compound is found.

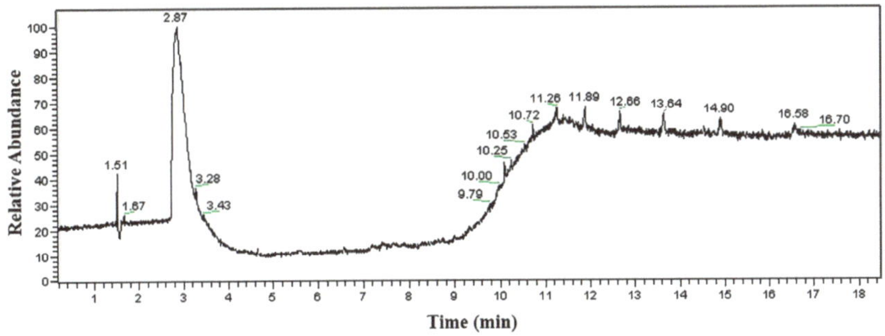


**Figure 10.** GC-MS analysis result of alkaline cleaning agent.

The analysis results of the neutral cleaning agent are shown in Figure 11. There are three obvious peaks in the spectrum, and two substances are identified. The organic compounds in the neutral cleaning agent mainly contain hydroxyl and ether groups, and the substances containing the ether group account for less.

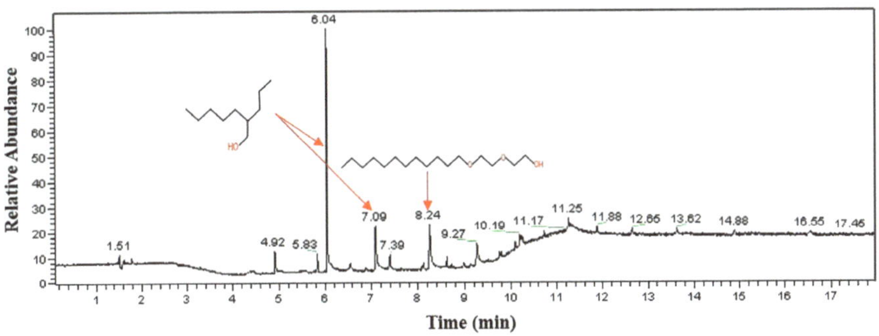


**Figure 11.** GC-MS analysis result of neutral cleaning agent.

The analysis results of the acidic cleaning agent are shown in Figure 12. The results indicate that the organic compounds in the acid cleaning agent also mainly contain hydroxyl and ether groups, and the substances containing the ether group account for less.

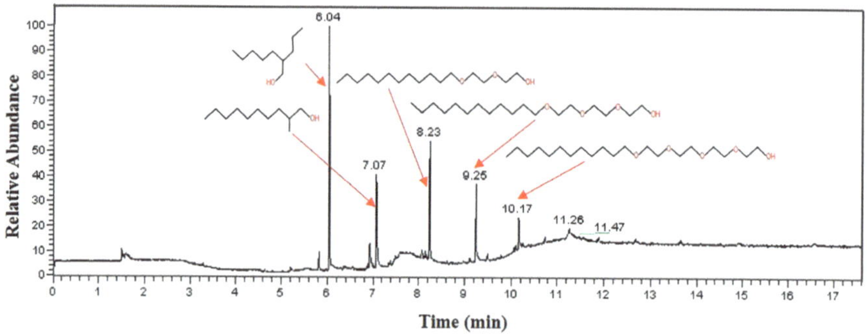


**Figure 12.** GC-MS analysis result of acidic cleaning agent.

The analysis results of the 602 cleaning agent are shown in Figure 13. The results prove that the main component of the 602 cleaning agent is 2-(2-n-butoxyethoxy) ethanol, accounting for 97.8% of the total content.

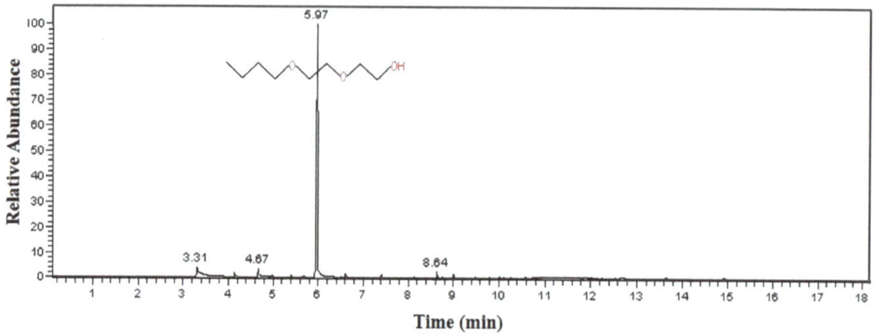


**Figure 13.** GC-MS analysis result of 602 cleaning agent.

### 3.3. Soaking Experiment

In order to better understand the cleaning and maintenance of the high-speed train, a field investigation was carried out. The field investigation shows that the high-speed train must be regularly cleaned and maintained, and four cleaning agents are used: acidic, neutral, alkaline, and 602 cleaning agents. The specific usage of the four cleaning agents is shown in Table 4. Since the acidic, neutral, and alkaline cleaning agents are sprayed on a large area during the cleaning process and the 602 cleaning agent is directly sprayed on the bogie with a high-pressure spray bottle, it is reasonable to speculate that these four cleaning agents would come into contact with the failed baffle.

**Table 4.** Specific usage of the four cleaning agents.

| Cleaning Agent | Main Ingredients | Concentration When Used | Cleaning Location | Cleaning Frequency |
|---|---|---|---|---|
| Alkaline | / | 20% | Below the blue line (Figure 1) | Four days |
| Neutral | Hydroxyl organics | 20% | Above the blue line (Figure 1) | Four days |
| Acid | Hydroxyl organics | 30% | All the body of the train | Three months |
| 602 | 2-(2-n-Butoxyethoxy) ethanol | 100% | Metal parts of the bogie | Overhaul |

To prove the conjecture that the failed baffle was severely corroded by the solvent, Sample 1# was cut into small pieces and soaked in different cleaning agents. The soaking results are shown in Figures 14 and 15. Figure 14 shows that after soaking in alkaline, neutral, and acidic cleaning agents, the thickness, mass, and hardness of the samples changed little (within 3%). Compared with the other three cleaning agents, the thickness and mass of the sample increased after soaking in the 602 cleaning agent, but the increase ratio is also small (within 3%). What is more, the hardness of the sample decreased significantly (about 12% after soaking for 5 days). Figure 15 shows that after soaking in the 602 cleaning agent for 6 days, the surface of the sample becomes powdery. This surface change is not observed after soaking in other solvents. The experimental results indicate that the 602 cleaning agent is the key factor in the failure of the baffle.

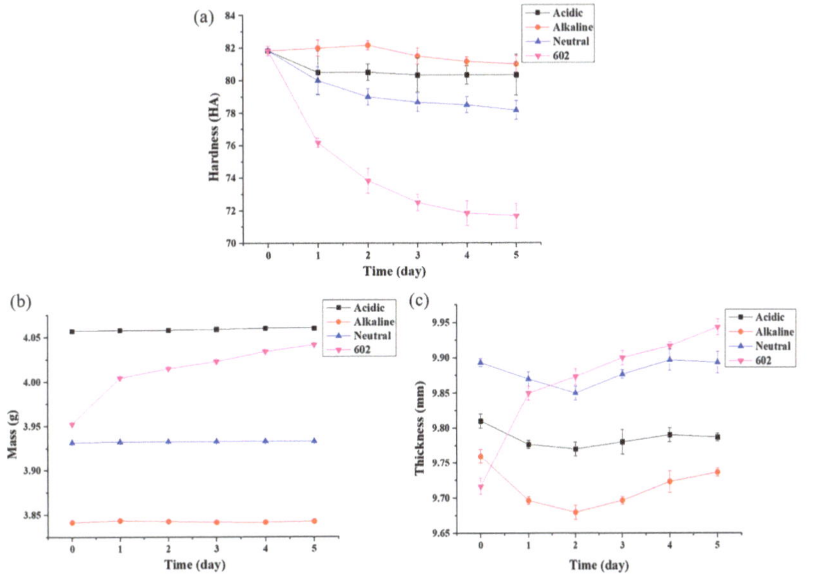


**Figure 14.** Change in hardness, mass, and thickness after soaking: (**a**) hardness; (**b**) mass; (**c**) thickness.

**Figure 15.** Surface changes after soaking in 602 cleaning agent.

In addition, according to the GC-MS test results, no organic matter is detected in the alkaline cleaning agent, but organic matter containing the ether group is detected in both the acidic and neutral cleaning agents. As the content of the organic matter containing the ether group in acidic and neutral cleaning agents is relatively small and has been diluted during use, the performance of the sample in acidic and neutral cleaning agents does not change significantly. The 602 cleaning agent contains 97.8% 2-(2-n-butoxyethoxy) ethanol and is used undiluted. This indicates that there is a higher concentration of organic matter containing the ether group in the 602 cleaning agent than in other cleaning agents. This is the reason why the performance of the sample in the 602 cleaning agent is seriously degraded.

## 4. Comprehensive Analysis

### 4.1. Environmental Stress Cracking

The performance test of the failed antenna baffle was carried out. The results of the FTIR test show that no obvious chemical changes occurred in the failed baffle, the results of the TGA test show that the heat resistance of the failed baffle decreases significantly, the results of the DSC test show that the Tg of the failed baffle decreases, and the results of the hardness test show that the surface hardness of the failed baffle decreases seriously. All these phenomena indicate that the failed baffle is seriously eroded by solvent. The soaking experiment results show that the hardness of the baffle decreases obviously after the baffle is exposed to the 602 cleaning agent, which further verifies that the failed baffle had the phenomenon of ESC.

ESC refers to the phenomenon in which polymers (especially glassy thermoplastics) are degraded by chemical agents in the presence of stress, resulting in damage to the polymer components. This is a solvent-induced failure as well as a material cracking that occurs in synergy between chemical agents and mechanical stress. Crystalline polymers and even lightly crosslinked polymers exhibit similar failures to glassy polymers, but generally require higher externally imposed stresses [24,25].

The reasons for ESC of the failed baffles can be explained by the following microscopic mechanism.

The molecular chain of PUR contains an isocyanate group (-NCO) and a urethane group (-NHCOO-), which are obtained by polymerization of isocyanate, polyols, and chain extenders. The chain extenders include small-molecule diols, diamines, and so on. The PUR molecular chain is composed of soft and hard segments. In PUR, polyester or polyether constitutes the soft segment, while isocyanate and the crosslinking agent constitute the hard segment, as shown in Figure 16 [26]. These two segments are thermodynamically incompatible. When the hard segments are close to each other, their electron orbitals easily overlap, resulting in the formation of hydrogen bonds between the hard segments. The hard–soft hydrogen bonds are easily formed between the secondary amino group and the soft segment. [27] These intermolecular hydrogen bonds make the hard segment embedded in the soft satin, which restricts the movement of the molecular chain of the soft segment, so that the Tg of the soft satin in PUR increases with the degree of hydrogen bonding, as illustrated in Figure 17a [28]. Because of its strong polarity, the hard-segment phase in PU forms a crystalline region through the interaction of intermolecular or intramolecular hydrogen bonds, which can play a role in physical crosslinking [29,30]. Due to the combined action of the hard segment as the physical crosslinking point and soft satin with high flexibility, PUR has the advantages of impact resistance, wear resistance, high elasticity, and high elongation [31,32]. Figure 17b shows that after soaking in the 602 cleaning agent, the PU rubber is swelled, and the molecular spacing increases, resulting in the rupture of hydrogen bonds. The rupture of the hydrogen bond results in the absence of physical crosslinking between hard segments, which leads to lower heat resistance, lower Tg, lower surface hardness, and the powdery surface of the PUR.

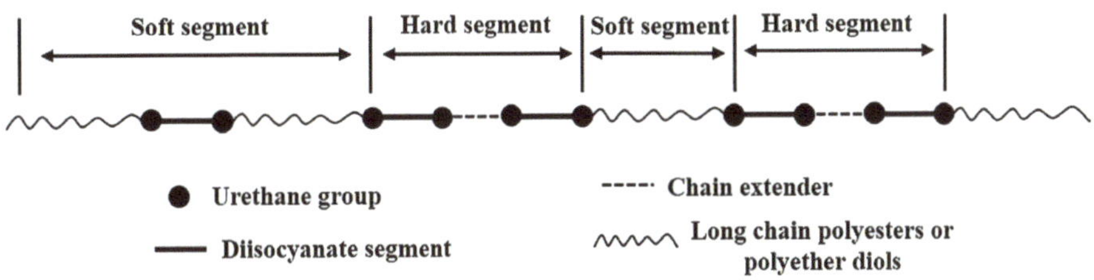


**Figure 16.** Molecular structure of PUR.

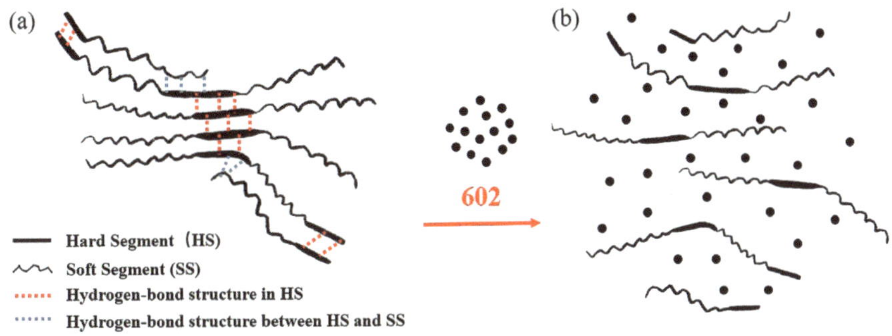


**Figure 17.** Changes in the internal structure of PUR before and after soaking: (**a**) internal structure of PUR before soaking; (**b**) internal structure of PUR after soaking.

The soaking experiment shows that the hardness of the antenna baffle decreases obviously after contacting the 602 cleaning agent. The microscopic mechanism reveals the mechanism of the decrease in the crosslinking density of PUR after soaking, which well explains the phenomenon of the surface hardness decrease and powdery surface of the baffle after soaking. More importantly, the soaking experiment and microscopic mechanism can perfectly explain the results of the Tg decrease, heat resistance decrease, crosslinking density decrease, hardness decrease, and powdery surface. This series of perfectly corroborating evidence indicated that the failed antenna baffle was exposed to the 602 cleaning agent. In addition, the high-speed train would be subjected to cyclic loads during running (as mentioned in Section 4.3 below). When solvent erosion and cyclic load are applied to the antenna baffle at the same time, ESC will occur, leading to a sharp decline in its mechanical properties. This will cause the antenna baffle to crack and fail in a short period of time. Therefore, ESC is the root cause of the premature failure of the antenna baffle.

### 4.2. Manufacturing Process

The new antenna baffle (Sample 1#) was observed under 3D-SM (Figure 18). Figure 18a shows that obvious bubbles can be observed on the surface of Sample 1#. The bubbles are large (visible to the naked eye) and abundant. In addition, at 160× magnification, by adjusting the focal length of the microscope to observe the same position, it can be found that different clear bubbles can be observed at different focal lengths (Figure 18c,d). This indicates that the bubbles are not at the same height, but distributed in different layers along the thickness direction.

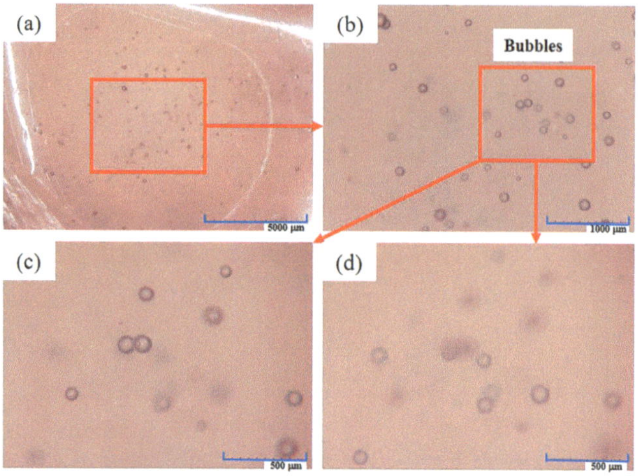


**Figure 18.** Surface appearance of Sample1# under 3D-SM: (**a**) overall appearance; (**b**) magnification; (**c**) first focal-length adjustment; (**d**) second focal-length adjustment.

The existence of bubbles will reduce the mechanical properties of the antenna baffle (it is easy to produce stress concentration at the bubbles), which is not conducive to its normal use. In addition, the presence of bubbles in the antenna baffle also makes it easier for the solvent to enter, and the solvent is more likely to erode the antenna baffle due to capillarity. Therefore, some bubbles left in the antenna baffle are an important cause of the premature failure of the antenna baffle.

### 4.3. Stress

During the normal operation of the train, because the antenna baffle is fixed at one end and free at the other, it is subjected to four different forces, which are inertia force, wind resistance, gravity, and vibration force caused by an uneven roadbed [33,34]. Since PUR has the characteristics of high hardness, good strength, high elasticity, and high wear and tear resistance, these four forces will not affect the use of the baffle under normal circumstances. However, when the baffle is eroded by the solvent, the physical properties of the baffle decrease significantly, and these four forces will provide the necessary external stresses for the ESC of the baffle. It is the combination of solvent erosion and the four forces that lead to ESC. Therefore, these four forces are another important cause of the premature cracking of the antenna baffle.

## 5. Conclusions and Recommendations

On the basis of the systematic failure analysis, the main conclusions could be drawn as follows:

(1) ATR-FTIR test results show that the material of the baffle is PU, and there is no obvious chemical change before and after the failure.

(2) NMR, TGA, DSC, and hardness test results show that compared with the new baffle, the crosslinking density, heat resistance, Tg, and hardness of the failed baffle are significantly reduced, indicating that the failed baffle was swelled by solvents. After soaking in the 602 cleaning agent for a period of time, it was found that the surface hardness of the new baffle decreased significantly. Therefore, during the cleaning and maintenance of the bogie, a small amount of the 602 cleaning agent was sprayed on the baffle, causing ESC. This is the root cause of premature cracking of the antenna baffle.

(3) There are some bubbles inside the antenna baffle when it is formed, which results in the decline of some of its mechanical properties. The presence of bubbles also causes

capillary action, making it easier for the cleaning agent to penetrate the baffle. This is an important cause of premature cracking of the antenna baffle.

(4) The antenna baffle withstands the forces caused by inertia, gravity, vibration force caused by uneven roadbed, and wind resistance. These four forces provide the necessary tensile and bending stresses for the cracking of the baffle. This is another important cause of premature cracking of the antenna baffle.

Based on the above conclusions, the following recommendations are put forward.

(1) When cleaning the bogie, it is necessary to cover the antenna baffle with a plastic film to prevent the 602 cleaning agent from being sprayed on the baffle. This can effectively avoid the swelling phenomenon.
(2) The supplier of antenna baffle products should improve the formulation process to ensure that the bubbles remaining in the baffle during the forming process are controlled within the acceptable quality range.
(3) The antenna baffle should avoid scratches, bumps, deformations, and other surface defects during transportation and storage. During installation, the surface state of the baffle should be carefully observed to avoid the use of defective baffles.

Finally, it is worth noting that the number of failed baffles was greatly reduced after the company adopted our solution, which proves that our recommendations are correct and effective.

**Author Contributions:** Investigation, T.-T.B.; Resources and writing—original draft preparation, C.S.; Supervision and project administration, Z.-G.Y. All authors have read and agreed to the published version of the manuscript.

**Funding:** This research received no external funding.

**Institutional Review Board Statement:** Not applicable.

**Informed Consent Statement:** Not applicable.

**Data Availability Statement:** Not applicable.

**Conflicts of Interest:** The authors declare no conflict of interest.

## References

1. Wang, S.Q.; Yuan, Y.; Tan, P.; Zhu, H.P.; Luo, K.T. Experimental study on mechanical properties of casting high-capacity polyurethane elastomeric bearings. *Constr. Build. Mater.* **2020**, *265*, 120725. [CrossRef]
2. Wypych, G. *Handbook of Polymers*; ChemTec Publishing: Toronto, ON, Canada, 2012; pp. 579–584.
3. Włoch, M.; Landowska, P. Preparation and Properties of Thermoplastic Polyurethane Composites Filled with Powdered Buckwheat Husks. *Materials* **2022**, *15*, 356. [CrossRef] [PubMed]
4. Alam, M.; Altaf, M.; Ahmed, M.; Shaik, M.R.; Wahab, R.; Shaik, J.P.; Samdani, M.S.; Ahmad, A. Development of Metallo (Calcium/Magnesium) Polyurethane Nanocomposites for Anti-Corrosive Applications. *Materials* **2022**, *15*, 8374. [CrossRef]
5. Uranbey, L.; Unal, H.I.; Calis, G.; Gumus, O.Y.; Katmer, S.; Karatas, C. One-pot preparation of electroactive shape memory polyurethane/carbon black blend. *J. Mater. Eng. Perform.* **2021**, *30*, 1665–1673. [CrossRef]
6. Jeong, H.; Ryu, S.; Kim, Y.N.; Ha, Y.M.; Tewari, C.; Kim, S.Y.; Kim, J.K.; Jung, Y.C. Easy, Fast Self-Heating Polyurethane Nanocomposite with the Introduction of Thermally Annealed Carbon Nanotubes Using Near-Infrared Lased Irradiation. *Materials* **2022**, *15*, 8463. [CrossRef] [PubMed]
7. Su, C.; Bi, T.T.; Pan, A.X.; Yi, G.; Luo, L.Y.; Chen, J.; Yang, Z.G. Failure analysis on abnormal cracking of polycarbonate plates used in the gearbox oil level gauge for high-speed train. *Eng. Fail. Anal.* **2023**, *143*, 106871. [CrossRef]
8. Ren, C.; Pan, H.L. An analysis of inclusion defects on strip steel surface induced by polyurethane roller. *Eng. Fail. Anal.* **2011**, *18*, 1122–1127. [CrossRef]
9. Junik, K.; Lesiuk, G.; Barcikowski, M.; Blazejewski, W.; Niemiec, A.; Grobelny, M.; Otczyk, K.; Correia, J.A.F.O. Impact of the hardness on the selected mechanical properties of rigid polyurethane elastomers commonly used in suspension systems. *Eng. Fail. Anal.* **2021**, *121*, 105201. [CrossRef]
10. Wen, M.C.; Tang, Y.Q.; Yuan, C.; Cui, E.H.; Ji, K.J.; Dai, Z.D. Effect of interconnected metal skeletons on the tribological properties of polyurethane elastomers. *Proc. Inst. Mech. Eng. Part J J. Eng. Tribol.* **2020**, *234*, 1635–1641. [CrossRef]
11. Cristiano, A.; Marcellan, A.; Keestra, B.J.; Steeman, P.; Creton, C. Fracture of model polyurethane elastomeric networks. *J. Polym. Sci. Part B Polym. Phys.* **2011**, *49*, 355–367. [CrossRef]
12. Mars, W.V.; Fatemi, A. A literature survey on fatigue analysis approaches for rubber. *Int. J. Fatigue* **2002**, *24*, 949–961. [CrossRef]

13. Tee, Y.L.; Loo, M.S.; Andriyana, A. Recent advances on fatigue of rubber after the literature survey by Mars and Fatemi in 2002 and 2004. *Int. J. Fatigue* **2018**, *110*, 115–129. [CrossRef]
14. Herzig, A.; Johlitz, M.; Lion, A. An experimental set-up to analyse the oxygen consumption of elastomers during ageing by using a differential oxygen analyser. *Contin. Mech. Thermodyn.* **2015**, *27*, 1009–1017. [CrossRef]
15. Haseeb, A.S.M.A.; Masjuki, H.H.; Siang, C.T.; Fazal, M.A. Compatibility of elastomers in palm biodiesel. *Renew. Energy* **2010**, *35*, 2356–2361. [CrossRef]
16. Mo, M.Y.; Gong, Y.; Yang, Z.G. Failure analysis on the O-ring of radial thrust bearing room of main pump in a nuclear power plant. *Eng. Fail. Anal.* **2020**, *115*, 104673. [CrossRef]
17. Su, C.; Pan, A.X.; Gong, Y.; Yang, Z.G. Failure analysis on rubber universal spherical joints for rail vehicles. *Eng. Fail. Anal.* **2021**, *126*, 105453. [CrossRef]
18. Loo, M.S.; Le Cam, J.B.; Andriyana, A.; Robin, E.; Afifi, A.M. Fatigue of swollen elastomers. *Int. J. Fatigue* **2015**, *74*, 132–141. [CrossRef]
19. Trakarnpruk, W.; Porntangjitlikit, S. Palm oil biodiesel synthesized with potassium loaded calcined hydrotalcite and effect of biodiesel blend on elastomer properties. *Renew. Energy* **2008**, *33*, 1558–1563. [CrossRef]
20. Gremlich, H.L. Infrared and Raman Spectroscopy. In *Handbook of Analytical Techniques*; John Wiley & Sons, Inc.: Hoboken, NJ, USA, 2008.
21. Kojio, K.; Nakashima, S.; Furukawa, M. Microphase-separated structure and mechanical properties of norbornane diisocyanate-based polyurethanes. *Polymer* **2007**, *48*, 997–1004. [CrossRef]
22. Kojio, K.; Mitsui, Y.; Furukawa, M. Synthesis and properties of highly hydrophilic polyurethane based on diisocyanate with ether group. *Polymer* **2009**, *50*, 3693–3697. [CrossRef]
23. Yang, X.L.; Wang, S.H.; Gong, Y.; Yang, Z.G. Effect of biological degradation by termites on the abnormal leakage of buried HDPE pipes. *Eng. Fail. Anal.* **2021**, *124*, 105367. [CrossRef]
24. Robeson, L.M. Environmental stress cracking: A review. *Polym. Eng. Sci.* **2013**, *53*, 453–467. [CrossRef]
25. Wright, C.D. *Environmental Stress Cracking of Plastics*; Lightning Source: La Vergne, TN, USA, 1996.
26. Mckeen, L.W. Elastomers and Rubbers. In *Permeability Properties of Plastics and Elastomers*, 4th ed.; Elsevier: Amsterdam, The Netherlands, 2017; pp. 209–247.
27. Steiner, T. The hydrogen bond in the solid state. *Angew. Chem. Int. Ed.* **2002**, *41*, 48–76. [CrossRef]
28. Scetta, G.; Selles, N.; Heuillet, P.; Ciccotti, M.; Creton, C. Cyclic fatigue failure of TPU using a crack propagation approach. *Polym. Test.* **2021**, *97*, 107140. [CrossRef]
29. Bonart, R.; Morbitzer, L.; Hentze, G. X-ray investigations concerning the physical structure of cross-linking in urethane elastomers. II. Butanediol as chain extender. *J. Macromol. Sci. Part B* **1969**, *3*, 337–356. [CrossRef]
30. Deblieck, R.A.C.; Van Beek, D.J.M.; Remerie, K.; Ward, I.M. Failure mechanisms in polyolefines: The role of crazing, shear yielding and the entanglement network. *Polymer* **2011**, *52*, 2979–2990. [CrossRef]
31. Hossieny, N.; Shaayegan, V.; Ameli, A.; Saniei, M.; Park, C.B. Characterization of hard-segment crystalline phase of thermoplastic polyurethane in the presence of butane and glycerol monosterate and its impact on mechanical property and microcellular morphology. *Polymer* **2017**, *112*, 208–218. [CrossRef]
32. Bhowmick, A.K.; Stephens, H. *Handbook of Elastomers*, 2nd ed.; CRC Press: Boca Raton, FL, USA, 2000.
33. Xiu, R.; Spiryagin, M.; Wu, Q.; Yang, S.; Liu, Y. Fatigue life assessment methods for railway vehicle bogie frames. *Eng. Fail. Anal.* **2020**, *116*, 104725. [CrossRef]
34. Xiu, R.; Spiryagin, M.; Wu, Q.; Yang, S.; Liu, Y. Fatigue life prediction for locomotive bogie frames using virtual prototype technique. *Proc. Inst. Mech. Eng. Part F J. Rail Rapid Transit* **2021**, *235*, 1122–1131. [CrossRef]

materials

MDPI

Article

# Condensation Flow Heat Transfer Characteristics of Stainless Steel and Copper Enhanced Tubes

**Xu Wang [1,*], David John Kukulka [2], Wei Li [3], Weiyu Tang [4] and Tianwen Li [3]**

1 College of Energy and Transportation Engineering, Inner Mongolia Agricultural University, 306 Zhaowuda Road, Hohhot 010018, China
2 Department of Mechanical Engineering Technology, State University of New York College at Buffalo, 1300 Elmwood Avenue, Buffalo, NY 14222, USA
3 Department of Energy Engineering, Zhejiang University, 38 Zheda Road, Hangzhou 310027, China
4 ZJU-Hangzhou Global Scientific and Technological Innovation Center, Hangzhou 311200, China
* Correspondence: yiqiyihui_wang@163.com

**Abstract:** In order to study the heat transfer of R410A in extreme environments, the properties of several stainless steel and copper-enhanced tubes were evaluated using R410A as the working fluid, and the results were compared with those of smooth tubes. Tubes evaluated include: smooth, herringbone (EHT-HB) and helix (EHT-HX) microgroove, herringbone/dimple (EHT-HB/D); herringbone/hydrophobic (EHT-HB/HY); and composite enhancement 1EHT (three-dimensional). Experimental conditions include a saturation temperature of 318.15K with a saturation pressure of 2733.5 kPa; a mass velocity in the range between 50 and 400 kg/(m$^2$·s); and an inlet quality controlled at 0.8 and an outlet quality of 0.2. Results indicate that the EHT-HB/D tube produces the best overall condensation heat transfer characteristics (high heat transfer performance and low frictional pressure drop). Using the performance factor (PF) to compare tubes for the range of conditions considered, the *PF* of the EHT-HB tube is greater than one, the *PF* of the EHT-HB/HY tube is slightly greater than one, and the *PF* of the EHT-HX tube is less than one. In general, as the mass flow rate increases, *PF* initially decreases and then increases. Previously reported smooth tube performance models that have been modified (for use with the EHT-HB/D tube) can predict the performance for 100% of the data points to within ±20%. Furthermore, it was determined that the thermal conductivity of the tube (when comparing stainless steel and copper) will have some effect on the tube-side thermal hydraulic performance. For smooth tubes, the heat transfer coefficients (HTC) of copper and stainless steel tubes are similar (with copper tube values being slightly higher). For enhanced tubes, performance trends are different; the HTC of the copper tube is larger than the SS tube.

**Keywords:** condensation heat transfer; enhanced tube; heat transfer coefficient; pressure drop

**Citation:** Wang, X.; Kukulka, D.J.; Li, W.; Tang, W.; Li, T. Condensation Flow Heat Transfer Characteristics of Stainless Steel and Copper Enhanced Tubes. *Materials* **2023**, *16*, 1962. https://doi.org/10.3390/ma16051962

Academic Editors: Andrea P. Reverberi and Balázs Illés

Received: 20 December 2022
Revised: 22 February 2023
Accepted: 22 February 2023
Published: 27 February 2023



## 1. Introduction

Passive enhanced heat transfer technology (i.e., enhanced surfaces) can significantly increase heat transfer with only a small pressure drop increase. Therefore, this is an important technology to consider in the field of heating, ventilation, and air conditioning (HVAC). Enhanced micro-fin tubes have been a topic of study for many researchers.

Micro-fin shape/form (including the various geometric structure parameters of the fin) are of interest; the effect of these parameters on heat transfer performance has been the topic of many studies. Kim et al. [1] examined the effect of micro-fin tube diameter on the boiling heat transfer characteristics of horizontal flow. As tube diameter increases, the HTC of micro-fin tubes is significantly higher than that of smooth tubes. Wellsandt and Vamling [2] found that the HTC of Y-shaped, micro-fin tubes is slightly larger than that of traditional spiral micro-fin tubes; however, there is a higher pressure drop in these tubes (when compared with smooth tubes). Wu et al. [3] experimentally studied the flow boiling of HTC in five different enhanced micro-fin heat exchange tubes. Experimental results [for mass flow rates less than 400 kg/(m$^2$·s)] show that micro-fin tubes with fin heights of 0.15 mm and fin apex angles of 25 and 30 degrees produce the best heat transfer performance. After studying the effect of the internal spiral angle on micro-fin tubes, Yang et al. [4] proposed an improved diagram that describes the internal flow pattern for horizontal micro-fin tubes; the transition curve to annular flow was determined to occur earlier in the micro-fin tube than what is found in the smooth tubes. Rollmann et al. [5] developed a novel pressure drop correlation (based on R407c and R410A pressure drop data) for micro-fin tubes.

With the recent development of enhanced three-dimensional tubes, studies of evaporation and condensation have become important topics to study. Li et al. [6] evaluated the condensation heat transfer characteristics of micro-fin tubes and compared them with smooth tubes. Kukulka et al. [7,8] experimentally evaluated changes in two-phase heat transfer for various enhanced tubes; they presented HTC and frictional pressure drop data for several enhanced three-dimensional tubes. Li et al. [9] compared the HTC of several stainless steel (SS) enhanced-surface tubes. Zhang et al. [10] carried out a similar analysis for a different set of conditions. Gu et al. [11] experimentally studied condensation heat transfer for moist air flows in enhanced tubes. Tang et al. [12] compared the condensation flow patterns found in a three-dimensional enhanced tube and detailed the transition conditions. Zhao et al. [13], Ali et al. [14], and Ji et al. [15] studied the effect that tube material thermal conductivity had on heat transfer of the fluid; Zhao et al. [13] studied carbon steel, copper, nickel, aluminum, and brass tube materials; Ali et al. [14] studied copper, brass, and bronze tube materials; and Ji et al. [15] discussed the relationship between fin thermal conductivity and fin efficiency.

Li et al. [16] studied the condensation of copper and SS-enhanced three-dimensional tubes using an R134a working medium. The HTC of the enhanced tube was 1.4~1.6 times higher than that of the smooth tube. A new correlation was presented for the enhanced tube that predicted the deviation of the condensation HTC in the tube to be $\pm$ 15%. Zheng et al. [17] carried out an experimental study on the evaporation characteristics of 410a in an enhanced three-dimensional SS tube. The largest evaporation HTC values were found in the EHT-HB/HY tube. Stratified wavy flow was found at the inlet of all tube types; the outlet flow pattern of the EHT-HB/HY and EHT-HB/D tubes was stratified wavy flow, while the outlet pattern of the EHT-HB tube was semi-annular flow, and for the EHT-HX tube it was annular flow. Kukulka et al. [18] presented the condensation heat transfer ratio (enhanced tube HTC/smooth tube HTC) of the enhanced tubes to be 1.15~2.05. Additionally, they found that the increase in heat transfer for enhanced tubes made of small tube diameters and/or high thermal conductivity materials was even larger. Some of the data used for comparison in this study came from Zheng et al. [17] and Kukulka et al. [18] and can be regarded as the continuation of those studies [16–18].

For the design, development, and assessment of high-performance heat transfer systems, it is important to study the heat transfer data of three-dimensional tubes. Additionally, the performance of enhanced tubes may or may not be better than that of smooth tubes. In order to determine how well enhanced tubes conduct heat and how material type may affect condensation heat transfer, an experimental study is necessary; additionally, in order to utilize enhancement results in industrial designs, an engineering model must be developed and validated. Smooth, herringbone (EHT-HB), and spiral (EHT-HX) microgrooves; herringbone/dimple (EHT-HB/D) herringbone/hydrophobic (EHT-HB/HY); and three-dimensional (composite dimple) 1EHT tubes are evaluated in the current study; their condensation heat transfer performance was analyzed and evaluated.

## 2. Experimental Details

Figure 1 presents the experimental setup, and Table 1 summarizes the physical parameters of tubes that were evaluated in this investigation. The first four tubes were used to study the influence of the enhancement structure on heat transfer, while the next three tubes were used to evaluate the effect of tube material on heat transfer, and, finally, the last three tubes were used to evaluate the effect of tube diameter. See [17] for additional experimental results for the first four tube types. See [18] for additional information on the other three tube types. See Figure 2 for an image of the enhancement surface of the evaluated tubes.

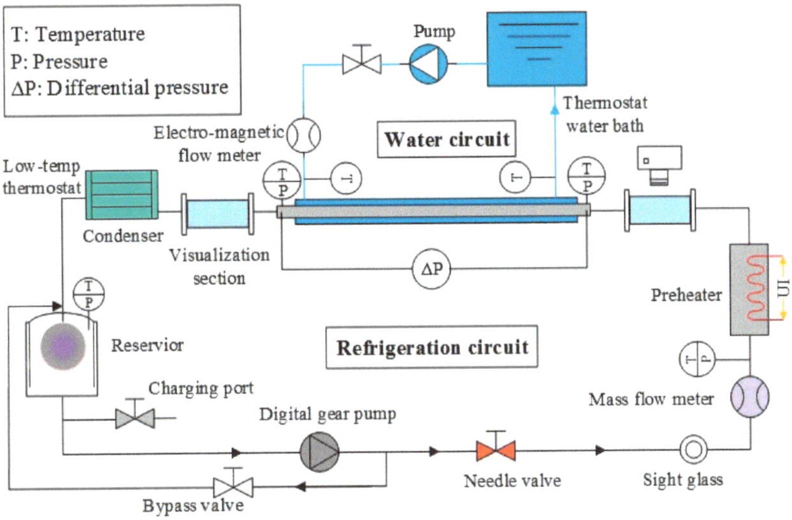


**Figure 1.** Experimental setup used in this study.

**Table 1.** Physical parameters of tubes evaluated in this study.

| Parameters | EHT-HX | EHT-HB | EHT-HB/D | EHT-HB/HY | Smooth | 1EHT | EHT-HX | EHT-HX | EHT-HX | EHT-HX |
|---|---|---|---|---|---|---|---|---|---|---|
| Material | SS | SS | SS | SS | Cu/SS | Cu/SS | Cu/SS | Al | Cu | Cu |
| Outer diameter (mm) | 12.7 | 12.7 | 12.7 | 12.7 | 9.52/12.7 | 9.52/12.7 | 9.52/12.7 | 7.0 | 7.0 | 7.0 |
| Thickness (mm) | 0.6 | 0.6 | 0.6 | 0.6 | 0.61 | 0.61 | 0.61 | 0.47 | 0.22 | 0.24 |
| Length (mm) | 2 | 2 | 2 | 2 | 2 | 2 | 2 | 2 | 2 | 2 |
| Dimple/fin height, mm | 1.14 | 0.08 | 1.21 | 0.08 | - | 0.19/1.71 | 0.25 | 0.06 | 0.10 | 0.15 |
| Dimple/fin pitch, mm | 5 | 0.8 | 4 | 0.8 | - | 0.35/1.34 | 0.31 | 0.32 | 0.32 | 0.35 |
| Dimple/fin width, mm | 2.3 | 0.31 | 3.51 | 0.31 | - | 4 | 0.8 | 0.07 | 0.06 | 0.17 |
| Helix angle, ° | 26.2 | 21 | 63 | 21 | - | 60 | 21 | 0 | 16 | 30 |

A horizontal test section was used with R410a refrigerant flowing on the inside of the tube and water flowing on the outside. Cooling water was measured using a flow meter; it then flowed through the test section and was then returned to the constant temperature water tank. Temperature was measured using a Pt100 platinum resistance temperature sensor. Refrigerant R410a was heated to a predetermined temperature and quality before it entered the test section; it was completely condensed and supercooled in the condenser; flow was measured using the refrigerant flow meter that was located in front of the preheating section and behind the booster pump; Pt100 platinum resistance sensors were installed at the inlet and outlet of preheating and test sections. A pressure gauge in the test section was used to measure the absolute pressure, along with a differential pressure gauge to measure the total differential pressure. There was a 20-channel data acquisition computer system that automatically recorded and stored all measured data.

Experimental conditions include: saturation temperature for condensation was 45 °C, with the mass flux adjusted according to the experimental needs [maximum of 500 kg/(m$^2$·s)], an inlet quality controlled at 0.8, and outlet quality maintained at 0.2.

Moffat [19] describes a method to calculate the uncertainty of directly measured and indirectly obtained parameters. Uncertainty in the parameters includes: uncertainty of the diameter was ±0.05 mm; length was ±0.5 mm; current was ±0.1 A; voltage was ±0.1 V;

temperature was ±0.1 K; uncertainty of pressure and differential pressure was ±0.075% of the full scale; flow uncertainty was ±0.2% of reading. Calculated parameter relative error of mass flux was ±1.18%; maximum relative error of quality was ±4.13%; and for the HTC, it was ±11.32%.

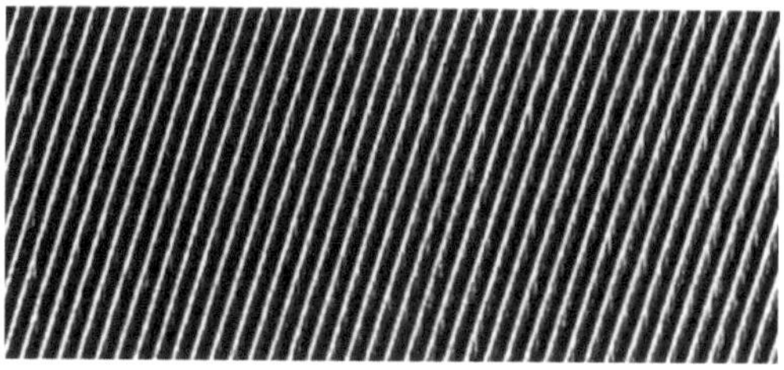

(**a**)

(**b**)

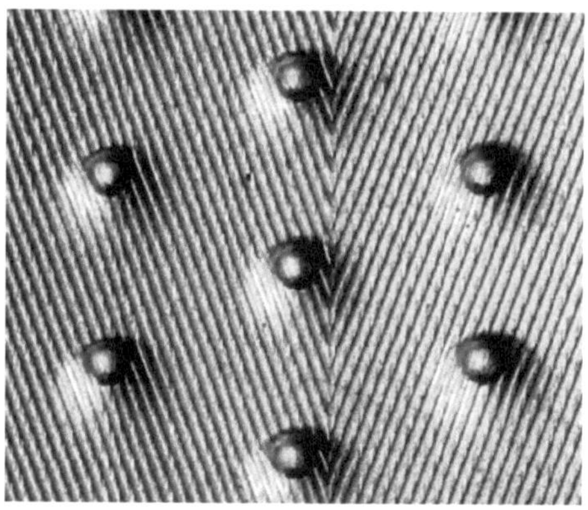

(**c**)

**Figure 2.** *Cont.*

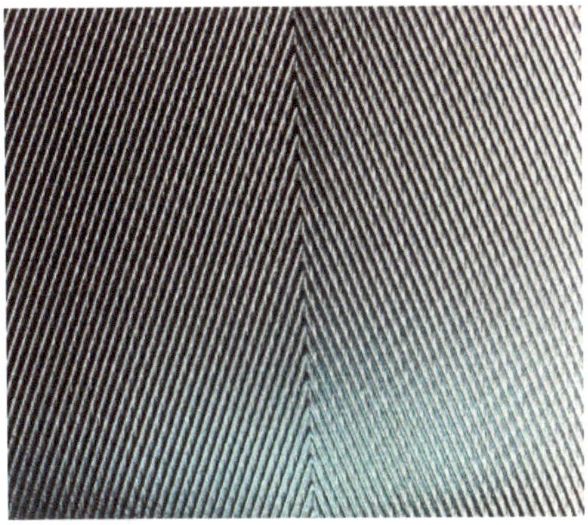

(**d**)

(**e**)

**Figure 2.** Surface images of the valued tubes: (**a**) EHT-HX; (**b**) EHT-HY; (**c**) EHT-HB/D; (**d**) EHT-HB; (**e**) 1EHT.

## 3. Results

### *3.1. Data Analysis*

Total heat transfer (see Equation (1)) is obtained from the heat balance of the water outside the tube,

$$Q_{t,ts} = C_{pl,w,ts} \times m_{w,ts} \times (T_{w,ts,in} - T_{w,ts,out}) \quad (1)$$

In Equation (1), $m_{w,ts}$ is the water flow in the test section; $c_{pl,w,ts}$ is the average specific heat capacity of water; $T_{w,ts,out}$ is the outlet temperature of water in the test section; and $T_{w,ts,in}$ is the inlet temperature of water in the test section.

Condensation HTC is obtained using Equations (2)–(7). The quality of the refrigerant in the inlet test section, $x_{in}$, is calculated from the heat exchange volume of the water in the preheating section, where the total heat transfer (see Equation (2)) of the refrigerant, $Q_{t,ph}$, consists of the liquid phase sensible heat (see Equation (3)) of the refrigerant, $Q_{sens}$, and the liquid-gas phase transition (see Equation (4)) latent heat, $Q_{lat}$.

$$Q_{t,ph} = C_{pl,w,ph} \times m_{w,ph} \times (T_{w,ph,in} - T_{w,ph,out}) = Q_{sens} + Q_{lat} \quad (2)$$

For Equation (2), $m_{w,ph}$ is the water flow in the preheating section; $c_{pl,w,ph}$ is the average specific heat capacity of the water; $T_{w,ph,out}$ is the outlet temperature of the water in the preheating section; and $T_{w,ph,in}$ is the inlet temperature of the water in the preheating section.

$$Q_{sens} = C_{pl,ref} \times m_{ref} \times (T_{sat} - T_{ref,ph,in}) \quad (3)$$

In Equation (3), $m_{ref}$ is the refrigerant mass flow rate; $c_{pl,ref}$ is the average specific heat capacity of the refrigerant; $T_{sat}$ is the refrigerant saturation temperature; and $T_{ref,ph,in}$ is the inlet temperature of the refrigerant in the preheating section.

$$Q_{lat} = m_{ref} \times h_{lv} \times x_{in} \quad (4)$$

For Equation (4), $h_{lv}$ is the latent heat of the refrigerant.

The quality in the outlet test section, $x_{out}$, is calculated from Equation (5).

$$x_{out} = x_{in} + Q_{t,ts}/(m_{ref} \times h_{lv}) \quad (5)$$

Inlet and outlet water temperatures and the saturated temperature of the refrigerant inside the tube are factors used to determine the logarithmic mean temperature difference (LMTD) [20,21] (see Equation (6)).

$$LMTD = \frac{(T_{w,ts,in} - T_{sat}) - (T_{w,ts,out} - T_{sat})}{\ln[(T_{w,ts,in} - T_{sat})/(T_{w,ts,out} - T_{sat})]} \quad (6)$$

Since the selected enhanced tubes being tested are brand new products that have not been used, the fouling thermal resistance can be ignored. Under the condition of ignoring fouling thermal resistance, the HTC of the inside tubes can be calculated using Equation (7).

$$h_{co} = \frac{1}{A_{ni}\left(\frac{Q_{t,ts}}{LMTD} - \frac{1}{A_0 h_0} - \frac{d_0 \ln\left(\frac{d_0}{d_i}\right)}{2k_{wall} A_0}\right)} \quad (7)$$

In Equation (7), $A_{ni}$ is the actual heat transfer area of the enhanced tubes; $A_0$ is the outer surface area of the evaluated tube; and $d_0$ is the outer diameter of the enhanced tube.

Previous research has demonstrated that the Gnielinski correlation [22] can be used to compute the turbulent, single-phase HTC for a smooth tube. Equation (8) presents the Gnielinski correlation that can be used to calculate the HTC of water on the outside of the smooth tube.

$$h_0 = \frac{(f/2)(Re - 1000)Pr}{1 + 12.7(f/2)^{1/2}(Pr^{2/3} - 1)} \left(\frac{\mu_{bulk}}{\mu_w}\right)^{0.14} \frac{k_w}{d_h} \quad (8)$$

Fanning friction (see Equation (9)) coefficient, $f$ is calculated using the Petukhov correlation [23] (applicable for smooth tubes in the Re range: $3000 < Re < 5 \times 10^6$).

$$f = (1.58\ln Re - 3.28)^{-2} \quad (9)$$

The surface enhancement of the inner and outer surfaces of the evaluated tube will affect the results. Therefore, the Gnielinski water-side HTC must be modified (using the Wilson graphic method) using the heat transfer enhancement coefficient $C$ (the ratio of the water-side HTC of the enhanced tube to the HTC of the smooth tube); the modified Gnielinski formula is given in Equation (10).

$$\frac{1}{Ch_0} = \frac{1}{U} - \frac{d_0}{d_i h_{ev}} - \frac{d_0 \ln(d_0/d_i)}{2k_{wall}} \quad (10)$$

In Equation (10), $U$ is the total HTC.

The friction pressure drop, $\Delta P_f$, is calculated using Equation (11).

$$\Delta P_f = \Delta P_t - \Delta P_g - \Delta P_m - \Delta P_{se} - \Delta P_{sc} \tag{11}$$

In Equation (11), $\Delta P_t$ is the total pressure drop; $\Delta P_g$ is the gravitational pressure drop; $\Delta P_m$ is the dynamic pressure drop; $\Delta P_{se}$ is the sudden expansion pressure drop; and $\Delta P_{sc}$ is the sudden contraction pressure drop. In this testing, all the tubes evaluated are placed horizontally, therefore, $\Delta P_g$ is equal to 0.

Equation (12) is used to calculate $\Delta P_m$.

$$\Delta P_m = G^2 \left\{ \left[ \frac{x}{\rho_v \varepsilon} - \frac{(1-x)^2}{\rho_l (1-\varepsilon)} \right]_{out} - \left[ \frac{x}{\rho_v \varepsilon} - \frac{(1-x)^2}{\rho_l (1-\varepsilon)} \right]_{in} \right\} \tag{12}$$

In Equation (12), $G$ is the mass flux rate; $x$ is the quality of the refrigerant; $\varepsilon$ is the cavitation rate; $\rho_v$ is the gas density of the refrigerant; $\rho_l$ is the density of the refrigerant liquid; $\varepsilon$ is calculated by Rouhani et al. [24] and shown in Equation (13).

$$\varepsilon = \frac{x}{\rho_v} \left\{ [1 + 0.12(1-x)] \left( \frac{x}{\rho_v} + \frac{1-x}{\rho_l} \right) + \frac{1.18(1-x)[g\sigma(\rho_l - \rho_v)^{0.25}]}{G\rho_l^{0.5}} \right\}^{-1} \tag{13}$$

Equation (14) can be used to calculate $\Delta P_{se}$ and Equation (15) is used to calculate $\Delta P_{sc}$.

$$\Delta P_{sc} = \frac{G^2}{2\rho_l} \left[ 1 - \left( \frac{\rho_l - \rho_v}{\rho_v} \right) \right] \tag{14}$$

$$\Delta P_{se} = \frac{G^2 \zeta (1-\zeta)}{\rho_l} \left[ 1 - \left( \frac{\rho_l - \rho_v}{\rho_v} \right) \right] \tag{15}$$

In Equation (15), $\zeta$ is the area ratio.

### 3.2. Evaluation of Smooth Tube Condensation Correlations

Shah et al. [25], Cavallini et al. [26], and Haraguchi et al. [27] present condensation correlations for smooth tubes. In this study, these three smooth tube correlations were compared in Figure 3 with experimental cavitation rate condensation results. The deviation of the HTC was predicted by Cavallini et al. [26] and is within ±9%. The deviation predicted by Shah et al. [25] and Haraguchi et al. [27] is in the range of −9% to +20%.

### 3.3. Condensation Enhanced Heat Transfer Factor in Enhanced Tubes

By introducing the heat transfer enhancement factor ($EF_h$), the heat transfer of the enhanced and smooth tube is compared (see Equation (16)). This factor is defined as the ratio of HTC of the enhanced tube to the HTC of an equal-diameter smooth tube under the same working conditions.

The $EF_h$ of composite surface-enhanced tubes (EHT-HB/D and EHT-HB/HY) are quite different; the $EF_h$ of an EHT-HB/D tube ranges from 1.4 to 1.75. This enhancement in the EHT-HB/D tube is the result of the surface producing an increase in the disturbance of the fluid; additionally, it also produces an increase in the turbulence intensity and improves the liquid drainage effect. The $EF_h$ of the EHT-HB/HY tube can reach 1.27; this is lower than that of the EHT-HB/D tube. This may be a result of the EHT-HB/HY tube fin structure; the HB/HY structure makes it difficult to remove the liquid at the hydrophobic fin and leads to the local heat transfer resistance increasing.

The heat transfer enhancement factors of the non-composite surface enhanced tubes (EHT-HX and EHT-HB) are also different; the $EF_h$ of the EHT-HB tube is higher than that of the EHT-HX tube. The fin structure in the EHT-HB tube makes it easier for the liquid to flow from the channel groove at the top of the fin; this creates a stronger disturbance in the flow process, thereby improving its HTC.

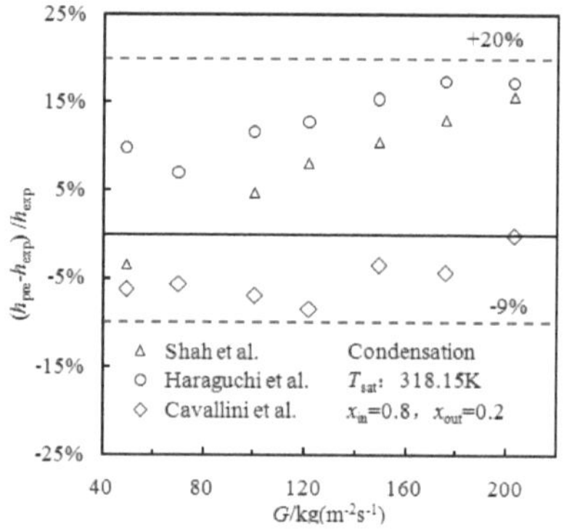


**Figure 3.** Comparison of experimental data and smooth tube condensation correlations.

$$EF_h = h_e/h_s \tag{16}$$

In Equation (16), $h_e$ is the HTC of enhanced tubes, and $h_s$ is the HTC of the smooth tube.

Figure 4 compares the $EF_h$ of the enhanced tubes. The data are from tubes 1 to 4 in Table 1. All the enhancement factors are greater than 1. At low flow rates, the trend of the $EF_h$ decreases, while for mass flow rates greater than 120 kg/(m$^2$·s), the $EF_h$ trend decreases slowly.

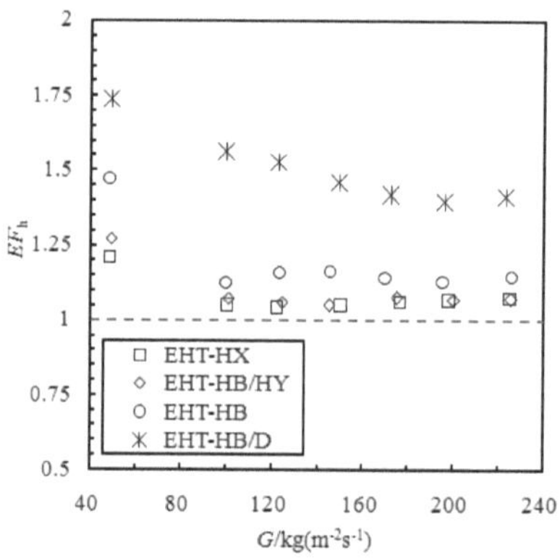


**Figure 4.** Comparison of enhancement factors of the evaluated enhanced tubes.

### 3.4. Evaluation of Enhanced Heat Transfer Performance Factor

The dimensionless parameter, performance factor (*PF*), is the enhanced heat transfer ratio (between enhanced and smooth tubes) divided by the pressure drop ratio. The following formula is used to calculate *PF*:

$$PF = (h_e/h_s) \times (p_s/p_e) \tag{17}$$

In Equation (17), $p_e$ is the frictional pressure drop of the enhanced tubes, and $p_s$ is the frictional pressure drop of the smooth tubes under the same working conditions.

Figure 5 presents the *PF* of the evaluated enhanced tubes. The data are from tubes 1 to 4 in Table 1. The performance factors of the EHT-HB/D and EHT-HB tubes are greater than 1; the *PF* of the EHT-HB/HY tube is slightly larger than 1; and finally, the *PF* of the EHT-HX tube is less than 1. The *PF* of the EHT-HB/D tube is the highest, reaching 1.3~1.5; this indicates that the surface structure can ensure high heat transfer performance and low frictional pressure drop. However, the *PF* of the EHT-HX tube is only 0.9~1.0, and the enhancement of heat transfer is not seen.

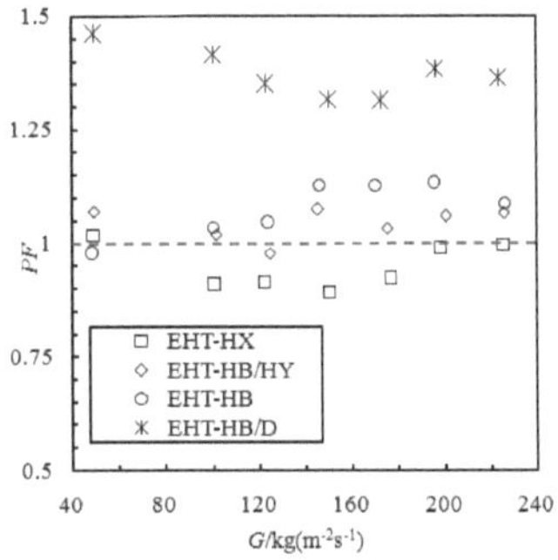


**Figure 5.** Comparison of performance factors of the evaluated enhanced tubes.

With increasing mass flow rate, the *PF* of the EHT-HB/D and EHT-HX tubes decreases initially and then increases, showing a parabolic shape in the low middle and higher values at both ends. The EHT-HB tube shows an initial rise, followed by a decrease; the curve shape is high in the middle and lower at the ends. Finally, the EHT-HB/D tube has the best condensation heat transfer performance, and the heat transfer of the EHT-HB/HY tube is not as good as that of the EHT-HB tube, while the EHT-HX tube has the worst performance factor curve.

### 3.5. Evaluation of Condensation Correlations in Enhanced Tubes

In previous condensation heat transfer studies, several researchers have proposed condensation heat transfer prediction models. However, the applicability of these models for use with enhanced heat transfer tubes remains to be verified. In this study, four models (Haraguchi et al. [27]; Huang et al. [28]; Chato [29]; Kedzierski and Goncalves [30]) are compared with the enhanced condensation data of this study. Results of that comparison are presented in Figure 6. The data are from tubes 1 to 4 in Table 1.

From Figure 6, it can be seen that the predicted trends of the four models for use in predicting the EHT-HX and EHT-HB/HY tubes are consistent (trends are also consistent with the trends from the previous EF analysis); this is the result of the $EF_h$ values for the EHT-HX and EHT-HB/HY tubes almost overlapping. Additionally, the $EF_h$ of the EHT-HB tube is slightly larger than that of the EHT-HX and EHT-HB/HY tubes; therefore, all four models produce the same trend for these types of tubes.

When considering the EHT-HX, EHT-HB/HY, and EHT-HB tubes, the deviation of all of the data points predicted by the Chato [29] model is within ±15%. For the EHT-HB/D tube, the deviation of all data points predicted by the Huang et al. [28] model is within ±30%, while for the Haraguchi et al. [27] model it is within ± 40%.

Since the HTC of the EHT-HB/D tube is the largest, the prediction accuracy of existing models is not ideal. It is noted that the Huang et al. [28] model can be seen as a modification of the Haraguchi et al. [27] model (see Table 2), with the Haraguchi et al. model data being based on R22, R134a, R123, etc.; additionally, the mass flow and Reynolds number cover a wide range. Therefore, it is necessary to modify the Haraguchi et al. model in order to make it applicable to the EHT-HB/D tube. The prediction results of the modified Haraguchi et al. model for the HB/D tube can be seen in Figure 7. Here, it can be seen that the modified Haraguchi et al. model can predict that 100% of the data points are within ±20%. .

### 3.6. Effect of Tube Material on Enhanced Heat Transfer Performance

Figure 8 compares the HTC as a function of mass flow rate for tubes (of the same diameter) produced from different materials. The data are from tubes 5 to 7 in Table 1. For smooth tubes, the HTC values of copper tubes and SS are similar (the copper tube HTC values are slightly higher). For enhanced tubes, the performance trends are different; the HTC of the copper tube is larger than the SS tube.

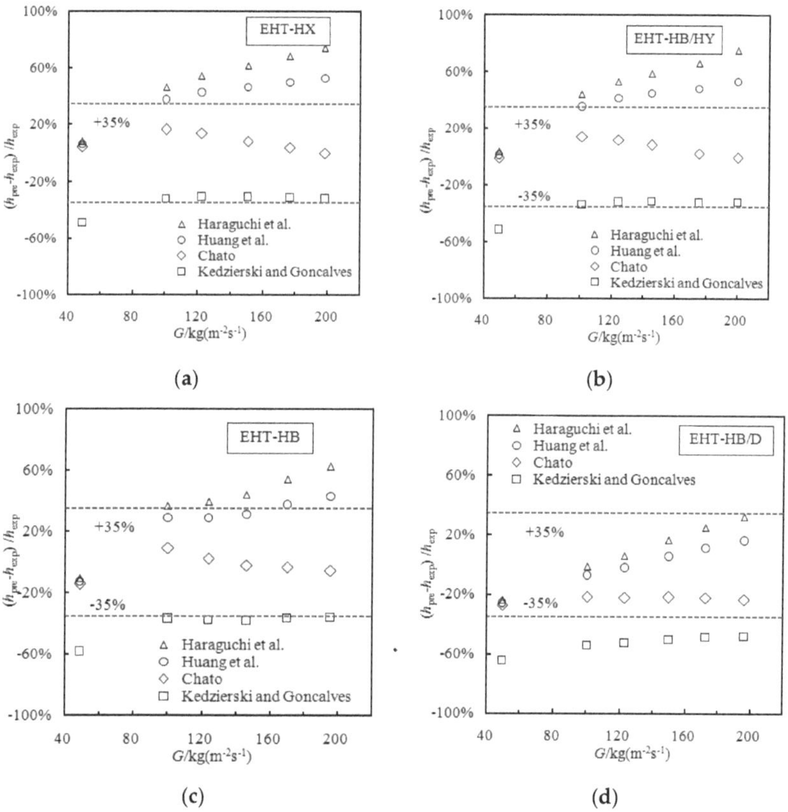


**Figure 6.** Comparison of using Haraguchi et al. [27]; Huang et al. [28]; Chato [29]; Kedzierski and Goncalves [30] correlations to predict performance for (**a**) EHT-HX, (**b**) EHT-HB/HY, (**c**) EHT-HB, and (**d**) EHT-HB/D.

**Table 2.** Modified Haraguchi et al. correlation.

| Author | Correlation |
| --- | --- |
| Modified Haraguchi et al. correlation | $h = \left(h_F^2 + h_B^2\right)^{\frac{1}{2}}$<br>$h_F = 0.0152\left(1 + 0.6Pr_l^{0.8}\right)\left(\frac{\Phi_V}{X_{tt}}\right)Re_l^{0.77}\frac{k_l}{d_h}$<br>$h_B = 0.790H(\varepsilon)\left(\frac{GaPr_l}{Ja_l}\right)^{\frac{1}{4}}\frac{k_l}{d_h}$<br>$\Phi_V = 1 + 0.5\left[\frac{G}{\sqrt{gd_h\rho_g(\rho_l-\rho_g)}}\right]^{0.75}X_{tt}^{0.35}$<br>$Ga = \frac{8\rho_l(\rho_l-\rho_g)d_h^3}{\mu_l^2}$<br>$H(\varepsilon) = \varepsilon + \left\{10\left[(1-\varepsilon)^{0.1} - 1\right] + 1.7 \times 10^{-4}Re_{lo}\right\}\sqrt{\varepsilon}\left(1 - \sqrt{\varepsilon}\right)$<br>$\varepsilon = \left[1 + \frac{\rho_g}{\rho_l}\left(\frac{1-x}{x}\right)\left(0.4 + 0.6\sqrt{\frac{\frac{\rho_l}{\rho_g}+0.4\left(\frac{1}{x}-1\right)}{1+0.4\left(\frac{1}{x}-1\right)}}\right)\right]^{-1}$<br>Applicable to the EHT-HB/D tube |

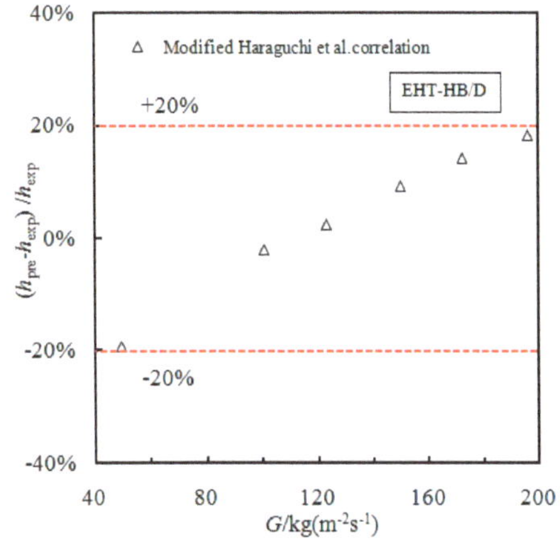


**Figure 7.** Comparison of the modified Haraguchi et al. correlation for use with the EHT-HB/D tube.

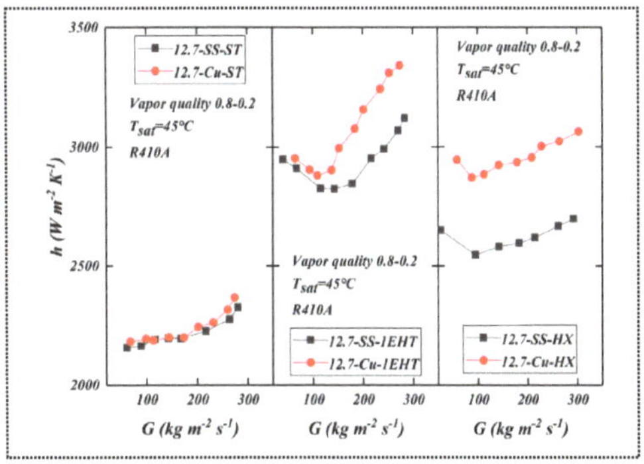


**Figure 8.** Comparison of HTC (as a function of mass flow rate) for tubes of the same diameter (12.7 mm) produced of different materials for smooth (ST) and enhanced (1EHT and HX) tubes.

Thermal conductivity varies between copper and SS; this affects the temperature distribution in the enhanced surface (fins); the higher the thermal conductivity of the material, the closer the temperature of the fin is to the temperature of the fin root. For low-conductivity materials, the opposite is true: the temperature of the fin deviates more from the temperature of the fin root. The temperature difference at different positions of the fins affects the efficiency. Fin efficiency is positively correlated with the thermal conductivity of the material, thus affecting the HTC. Therefore, the HTC of copper tube is higher than that of SS tube. It can also be seen from the figure that the influence of tube material composition on the HTC of the EHT-HX tube is larger; the tube produced from copper is significantly higher than SS tubes. Differences in performance between enhancement types are also noted; however, this is a function of differences in the area enhancement ratio between the EHT-HX and 1EHT tubes; additionally, there are differences in the enhancement characters used. The former has a higher area ratio than the latter; therefore, temperature difference and fin efficiency have a greater effect on performance.

For smooth tubes, the HTC increases with the increase in mass flow rate, and in the middle region, the increase is relatively slow. However, for the enhanced tube, the HTC initially decreases and then increases; they reach their minimum values near 100 kg/(m$^2$·s). This phenomenon is related to flow pattern, shear force, surface tension, etc., and requires further research.

Figure 9 shows the comparison of heat transfer resistance values (copper and SS) that take place during condensation when using enhanced tubes. The data are from tubes 5 to 7 in Table 1. From Figure 9, it can be concluded that the thermal resistance of the tube material is the least important component when looking at the total thermal resistance of the tube; however, the total thermal resistance is still an important component. The thermal resistance ratio of SS tube is higher than that of copper tube; additionally, the 1EHT tube is enhanced on both sides (inside and outside), therefore its external heat resistance is lower than that of either the smooth tube or the EHT-HX tube.

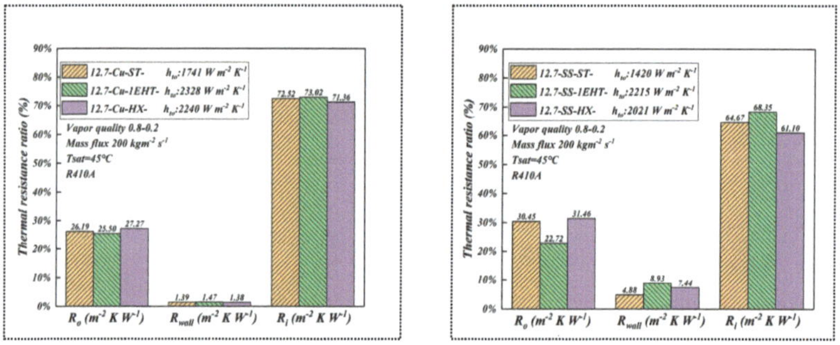


(a) (b)

**Figure 9.** Comparison of condensation thermal resistance ratio of different tube materials (**a**) copper (**b**) SS.

According to Figure 10, it can be seen that the tube diameter and material will affect the HTC. The data are from tubes 5 to 7 in Table 1. Larger values of the HTC in the 9.52 mm tube are seen when compared with the HTC values in the 12.7 mm tube. The HTCs of the enhanced tubes initially decrease and then increase; however, the turning point for the small tube diameter is delayed. Gravity is playing a leading role in heat transfer that takes place in the tube with the larger diameter, while in the smaller tube diameter, shear force and surface tension play the leading roles; this is conducive to the uniform distribution of the liquid film on the tube diameter. Additionally, in the smaller tube diameter, there is a larger heat flux per unit volume; all these factors contribute to the larger HTC that is shown in the enhanced 9.52 mm tube (when compared with the HTC of the 12.7 mm tube).

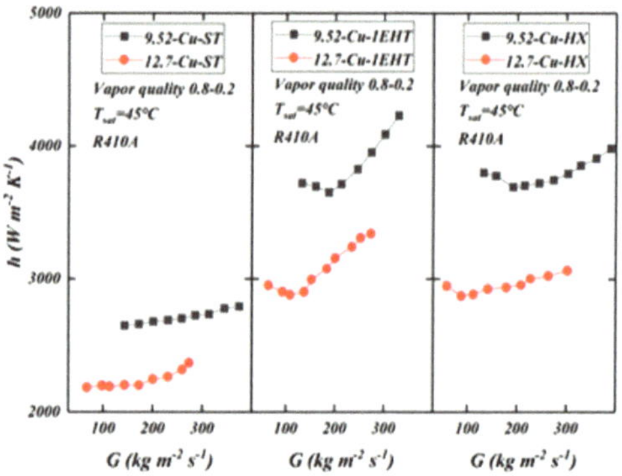


**Figure 10.** Comparison of HTC (as a function of mass flow rate) for different diameter (9.52 and 12.7 mm) tubes produced from copper, for smooth (ST) and enhanced (1EHT and HX) tubes.

### 3.7. Influence of Tube Diameter on Condensation HTC

Figure 11 shows the effect of tube diameter on condensation HTC. The data are from tubes 5 to 10 in Table 1. It can be seen that with the decrease in tube diameter, the condensation HTC in the enhanced tube shows an increasing trend; additionally, the tube diameter has a greater impact on the condensation HTC than the enhanced surface structure. This can be explained as follows: with the decrease in tube diameter, shear force and surface tension gradually replace gravity and become the dominant forces. This is beneficial to remove and dilute the liquid film at the bottom. The smaller the tube diameter, the higher the surface area density (surface area volume ratio); this results in a higher heat flux per unit volume.

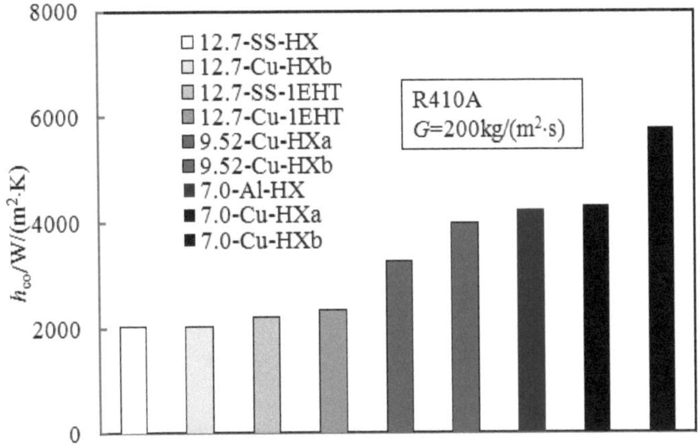


**Figure 11.** Effect of tube diameter on condensation HTC.

## 4. Conclusions

Through experimental methods, the condensation of R410A in different enhanced tubes was studied, and HTC and pressure drop data were obtained. Through a series of analyses, the following conclusions were obtained:

(1) The $EF_h$ of the EHT-HB/D tube is the highest; its performance is closely related to increasing fluid disturbance and improving drainage. The structure of the EHT-HB/HY tube increases the local thermal resistance and inhibits heat transfer, while the drainage effect of the EHT-HB tube is better than that of the EHT-HX tube.

(2) The best overall condensation heat transfer resistance characteristics (and the highest PF) are shown in the EHT-HB/D tube; it has a low friction pressure drop and high heat transfer performance. Additionally, the *PF* of the EHT-HB tube is greater than one, the *PF* of EHT-HB/HY is slightly higher than one, and the *PF* of EHT-HX is less than one. In general, when the mass flow rate increases gradually, the *PF* initially decreases and then increases.

(3) Correlations that predict the condensation HTC of enhanced tubes are discussed and modified. For EHT-HX, EHT-HB/HY, and EHT-HB tubes, the deviation of all data points predicted by the Chato model is within ±15%. However, for the EHT-HB/D tube, a modification to the Haraguchi et al. model is necessary, and that modification can predict that 100% of the data points are within ±20%.

(4) Thermal conductivity (SS and copper) of the tube material for smooth tubes has a minimal effect on its thermal hydraulic performance; however, for the enhanced tubes, the HTC increases with increasing thermal conductivity in extreme environments of condensation flow, at the saturation temperature of 318.15K with a saturation pressure of 2733.5 kPa. The refrigerant side convective heat transfer resistance dominates, while the wall heat transfer resistance makes up a modest fraction of the overall resistance. The thermal conductivity of the enhancement character (fin) affects the heat transfer for enhanced tubes.

(5) The effect of tube diameter on condensation heat transfer in the enhanced tubes with different tube diameters is much higher than that of the enhancement surface structure.

With the reduction of tube diameter, shear force and surface tension gradually replace gravity and become the dominant forces; this is conducive to the removal of the liquid film at the bottom.

Future research will continue this study and also provide a simulation of this work.

**Author Contributions:** Conceptualization, W.L.; methodology, W.T.; validation, T.L.; investigation, W.T.; resources, W.L.; data curation, T.L.; writing—original draft preparation, X.W.; writing—review and editing, W.L. and D.J.K.; supervision, D.J.K.; project administration, W.L. All authors have read and agreed to the published version of the manuscript.

**Funding:** This work is supported by the National Science Foundation of China (52076187).

**Institutional Review Board Statement:** Not applicable.

**Informed Consent Statement:** Not applicable.

**Data Availability Statement:** The pre-processed data used in this study are available on request from the corresponding author.

**Acknowledgments:** We thank Jincai Du, the engineer, for his guidance and help with the experimental device.

**Conflicts of Interest:** The authors declare no conflict of interest.

## Nomenclature

| | |
|---|---|
| $A$ | test tube surface area, $m^2$ |
| $C$ | enhancement ratio |
| $c_p$ | specific heat, J/(kg·K) |
| $D$ | dimple |
| $d$ | test tube diameter, m |
| $d_h$ | hydraulic diameter, m |
| $Fa$ | $(\rho_l-\rho_v)\,\sigma/G^2D_h$ |
| $f$ | Fanning friction factor |
| $G$ | mass flux, $kg/(m^2{\cdot}s)$ |
| $Ga$ | Galileo number |
| $g$ | gravitational acceleration, $m/s^2$ |
| HB | herringbone |
| HB/D | herringbone dimple |
| HB/HY | hydrophobic herringbone |
| HX | spiral microgrooves |
| $h$ | heat transfer coefficient, $W/(m^2{\cdot}K)$ |
| $h_{lv}$ | latent heat of vaporization, J/kg |
| $k$ | thermal conductivity, W/(m·K) |
| $Ja$ | Jakob number |
| $l$ | liquid only |
| $L$ | tube length, m |
| $LMTD$ | logarithmic mean temperature, K |
| $m$ | mass flux, kg/s |
| $M$ | molecular weight |
| $P$ | Pressure, kPa |
| $PF$ | performance factor |
| $Pr$ | Prandtl number |
| $Q$ | heat transfer amount, W |
| $q$ | heat flux, $W/m^2$ |
| $Re$ | Reynolds number |
| $T/t$ | temperature, K/°C |
| $U$ | Total heat transfer coefficient, $W/(m^2{\cdot}K)$ |
| $x$ | vapor quality |
| $X_{tt}$ | Martinelli parameter $X_{tt}=\left(\frac{1-x}{x}\right)^{0.9}\left(\frac{\rho_v}{\rho_l}\right)^{0.5}\left(\frac{\mu_l}{\mu_v}\right)^{0.1}$ |

| | |
|---|---|
| *Greek symbols* | |
| $\mu$ | dynamic viscosity, Pa·s |
| $\rho$ | density, $kg/m^3$ |
| $\varepsilon$ | void fraction |
| $\sigma$ | surface tension, N/m |
| $\zeta$ | area ratio |
| $\Phi$ | two-phase multiplier |
| *Subscripts* | |
| *bulk* | bulk temperature |
| *exp* | experimental |
| *f* | frictional |
| *g* | gravitational |
| *i* | inner |
| *in* | inlet |
| *l* | liquid phase |
| *lat* | latent heat |
| *m* | momentum |
| *ni* | actual heat transfer area |
| *o* | outer |
| *out* | outlet |
| *ph* | preheating section |
| *pre* | predictive |
| *ref* | refrigerant |
| *s* | smooth |
| *sat* | saturated |
| *sc* | sudden contraction |
| *se* | sudden enlargement |
| *sens* | sensible heat |
| *t* | total |
| *te* | test section |
| *tp* | two-phase |
| *ts* | test section |
| *v* | vapor phase |
| *wall* | wall parameters |
| *w* | water |

## References


1. Kim, Y.; Seo, K.; Chung, J.T. Evaporation heat transfer characteristics of R-410A in 7 and 9.52 mm smooth/micro-fin tubes. *Int. J. Refrig.* **2002**, *25*, 716–730. [CrossRef]
2. Wellsandt, S.; Vamling, L. Evaporation of R134a in a horizontal herringbone microfin tube: Heat transfer and pressure drop. *Int. J. Refrig.* **2005**, *28*, 889–900. [CrossRef]
3. Wu, Z.; Wu, Y.; Sundén, B.; Li, W. Convective vaporization in micro-fin tubes of different geometries. *Exp. Therm. Fluid Sci.* **2013**, *44*, 398–408. [CrossRef]
4. Yang, C.M.; Hrnjak, P. A new flow pattern map for flow boiling of R410A in horizontal micro-fin tubes considering the effect of the helix angle. *Int. J. Refrig.* **2020**, *109*, 154–160. [CrossRef]
5. Rollmann, P.; Spindler, K. New models for heat transfer and pressure drop during flow boiling of R407C and R410A in a horizontal microfin tube. *Int. J. Therm. Sci.* **2016**, *103*, 57–66. [CrossRef]
6. Li, G.; Huang, L.; Tao, L. Experimental investigation of refrigerant condensation heat transfer characteristics in the horizontal microfin tubes. *Appl. Therm. Eng.* **2017**, *123*, 1484–1493. [CrossRef]
7. Kukulka, D.J.; Smith, R.; Li, W. Condensation and Evaporation Characteristics of Vipertex 1EHT Enhanced Heat Transfer Tubes. *Chem. Eng. Trans.* **2014**, *39*, 727–732.
8. Kukulka, D.J.; Smith, R.; Li, W. Experimental Comparison of the Evaporation and Condensation Heat Transfer Coefficients on the Outside of Enhanced Surface Tubes with Different Outer Diameters. *Chem. Eng. Trans.* **2019**, *76*, 31–36.
9. Li, W.; Tang, W.; Gu, Z.; Guo, Y.; Ma, X.; Minkowycz, W.J.; Kukulka, D.J. Analysis of condensation and evaporation heat transfer inside 3-D enhanced tubes. *Numer. Heat Transf. Appl.* **2020**, *78*, 525–540. [CrossRef]
10. Zhang, J.; Zhou, N.; Li, W.; Luo, Y.; Li, S. An experimental study of R410A condensation heat transfer and pressure drops characteristics in microfin and smooth tubes with 5 mm OD. *Int. J. Heat Mass Transf.* **2018**, *125*, 1284–1295.

11. Gu, Y.H.; Liao, Q.; Cheng, M.; Ding, Y.D.; Zhu, X. Condensation heat transfer characteristics of moist air outside a three-dimensional finned tube. *Int. J. Heat Mass Transf.* **2020**, *158*, 119983. [CrossRef]
12. Tang, W.; Sun, Z.C.; Li, W. Visualization of flow patterns during condensation in dimpled surface tubes of different materials. *Int. J. Heat Mass Transf.* **2020**, *161*, 120251. [CrossRef]
13. Zhao, C.Y.; Ji, W.T.; Jin, P.H.; Zhong, Y.J.; Tao, W.Q. The influence of surface structure and thermal conductivity of the tube on the condensation heat transfer of R134a and R404A over single horizontal enhanced tubes. *Appl. Therm. Eng.* **2017**, *125*, 1114–1122.
14. Ali, H.M.; Briggs, A. Condensation heat transfer on pin-fin tubes: Effect of thermal conductivity and pin height. *Appl. Therm. Eng.* **2013**, *60*, 465–471. [CrossRef]
15. Ji, W.T.; Zhao, C.Y.; Zhang, D.C.; Li, Z.Y.; He, Y.L.; Tao, W. Q Condensation of R134a outside single horizontal titanium, cupronickel (B10 and B30), stainless steel and copper tubes. *Int. J. Heat Mass Transf.* **2014**, *77*, 194–201. [CrossRef]
16. Li, W.; Guo, Y.; Gu, Z.B.; Ma, X.; Ayub, Z.; He, Y.; Kukulka, D.J. An Experimental Study of R134a Condensation Heat Transfer in Horizontal Smooth and Enhanced Tubes. *J. Heat Transf.* **2020**, *142*, 071603. [CrossRef]
17. Zheng, B.; Wang, J.; Guo, Y.; Kukulka, D.J.; Tang, W.; Smith, R.; Sun, Z.; Li, W. An Experimental Study of In-Tube Condensation and Evaporation Using Enhanced Heat Transfer (EHT) Tubes. *Energies* **2021**, *14*, 867. [CrossRef]
18. Kukulka, D.J.; Li, W.; Smith, R. An Experimental Investigation to Determine the Effect of Tube Material on the Tubeside Heat Transfer Performance of the Enhanced 1EHT Three Dimensional Heat Transfer Tube. *Chem. Eng. Trans.* **2021**, *88*, 397–402.
19. Moffat, R.J. Describing the uncertainties in experimental results. *Exp. Therm. Fluid Sci.* **1988**, *1*, 3–17. [CrossRef]
20. Ram, P. An Improved Approved Approximation for the Log-Mean Temperature Difference. *Chem. Eng.* **1988**, *95*, 110.
21. Ziegler, F. Relationships between Temperature Differences in Heat Exchangers of Heat Transformation Devices. *Rev. Gen. De Therm.* **1998**, *37*, 549–555. [CrossRef]
22. Gnielinski, V. New Equations for Heat and Mass Transfer in Turbulent Pipe and Channel Flow. *Int. Chem. Eng.* **1976**, *16*, 8–16.
23. Petukhov, B.S. Heat transfer and friction in turbulent pipe flow with variable physical properties. *Adv. Heat Transf.* **1970**, *6*, 503–564.
24. Rouhani, S.Z.; Axelsson, E. Calculation of void volume fraction in the subcooledand quality boiling regions. *Int. J. Heat Mass Transf.* **1970**, *13*, 383–393. [CrossRef]
25. Shah, M.M. A general correlation for heat transfer during film condensation inside pipes. *Int. J. Heat Mass Transf.* **1979**, *22*, 547–556. [CrossRef]
26. Cavallini, A.; Col, D.D.; Doretti, L.; Matkovic, M.; Rossetto, L.; Zilio, C.; Censi, G. Condensation in Horizontal Smooth Tubes: A New Heat Transfer Model for Heat Exchanger Design. *Heat Transf. Eng.* **2006**, *27*, 31–38. [CrossRef]
27. Haraguchi, H.; Koyama, S.; Fujii, T. Condensation of refrigerants HCFC22, HFC134a and HCFC123 in a horizontal smooth tube:2nd Report, Proposals of Empirical Expressions for Local Heat Transfer Coefficient. *J. Jpn. Soc. Mech. Eng.* **1994**, *60*, 2117–2124.
28. Huang, X.; Ding, G.; Hu, H.; Zhu, Y.; Peng, H.; Gao, Y.; Deng, B. Influence of oil on flow condensation heat transfer of R410A inside 4.18 mm and 1.6 mm inner diameter horizontal smooth tubes. *Int. J. Refrig.* **2010**, *33*, 158–169. [CrossRef]
29. Chato, J. Laminar Condensation of Inside Horizontal and Inclined Tubes. *Am. Soc. Heat. Refrig. Air Cond. Eng. J.* **1962**, *4*, 52–60.
30. Kedzierski, M.A.; Goncalves, J.M. Horizontal Convective Condensation of Alternative Refrigerants within a Micro-Fin Tube. *J. Enhanc. Heat Transf.* **1999**, *6*, 161–178. [CrossRef]

materials

Article

# Research on Temperature Field of Controllable Bonded Prestressed Structure Based on Electrothermal Method

Xueyu Xiong [1,2,*] and Nan Jiang [1]

1 Department of Structural Engineering, College of Civil Engineering, Tongji University, Shanghai 200092, China; j_nan@tongji.edu.cn
2 Key Laboratory of Advanced Civil Engineering Materials, Tongji University, Shanghai 200092, China
* Correspondence: xiong_xueyu@tongji.edu.cn

**Abstract:** Controllable bonded prestress represents an innovative advancement stemming from retard-bonded prestress systems, distinguished by its intrinsic controllability in bonding. The controllable bonded binder can be artificially heated and cured rapidly through DC heating after the completion of prestressed tension, allowing for enhanced control over the process. FLUENT simulates controllable bonded prestressed structure's temperature field, yielding a 1.73% max error validated against measured data. Based on the theory of heat transfer, the maximum error of the calculated temperature field of the controllable bonded test beam under DC heating using the Stehfest numerical algorithm is 1.28%, which exhibits a strong alignment with both simulated and measured results. The parameter analysis identifies current, binder thickness, and steel-strand diameter as key temperature distribution influencers. The relationship between the current and heating time follows a quadratic inverse pattern. Increasing the heating current can significantly reduce the duration of heating. Under identical heating conditions, the temperature of the controllable binder is directly proportional to its thickness. A higher thickness results in a higher temperature. Additionally, larger diameters of steel-stranded wire lead to a lower heating efficiency.

**Keywords:** controllable bonded prestressed concrete; electrothermal method; experimental verification; numerical simulation; temperature field

**Citation:** Xiong, X.; Jiang, N. Research on Temperature Field of Controllable Bonded Prestressed Structure Based on Electrothermal Method. *Materials* **2023**, *16*, 7108. https://doi.org/10.3390/ma16227108

Academic Editor: Carlos Leiva

Received: 24 September 2023
Revised: 29 October 2023
Accepted: 31 October 2023
Published: 9 November 2023



## 1. Introduction

Ensuring complete compaction during grouting poses challenges in maintaining construction quality for bonded post-tensioned pre-stressed concrete, giving rise to a cascade of engineering issues [1]. In response, Japan pioneered the development of slow-bonded prestressed reinforcement in the 1980s, circumventing the need for concrete grouting [2]. Initially, minimal bonding force exists between the prestressed reinforcement and the slow-bonded material, resembling an unbonded system. Subsequent curing of the slow-bonded material after late tensioning achieves bonding system equivalence. A central component of the retard-bonded prestress system, the binder, typically comprises a gradually solidifying thermosetting material from its production stage. The standard tension application window spans 60–240 days, with a corresponding curing period of 180–720 days [3–5]. This protracted curing timeframe, coupled with incompatibility with dynamic loads during curing, significantly constrains its application in bridge and railway engineering.

Controllable bonded prestressed reinforcement denotes prestressed reinforcement subject to artificial control over the slow-setting adhesive's curing process. Before tensioning, the binder experiences gradual or even no curing, transitioning to rapid curing and adhesive properties post-tensioning, thereby aligning more effectively with engineering requirements [6]. The controllable bonded tendon closely resembles the retard-bonded tendon, comprising a steel strand, controllable bonded binder, and duct. The pivotal controllable bonded binder chiefly comprises epoxy resin, latent heat-sensitive curing agent, and diverse modified materials. Pre-tensioning exists in a semi-flow state, permitting free sliding of the steel strand within the duct. Upon prestressed tendon elongation, controllable heating of the adhesive tendon triggers latent curing agent activation, promoting rapid curing upon reaction with other components. Figure 1 displays the profile construction of controllable bonded strands.

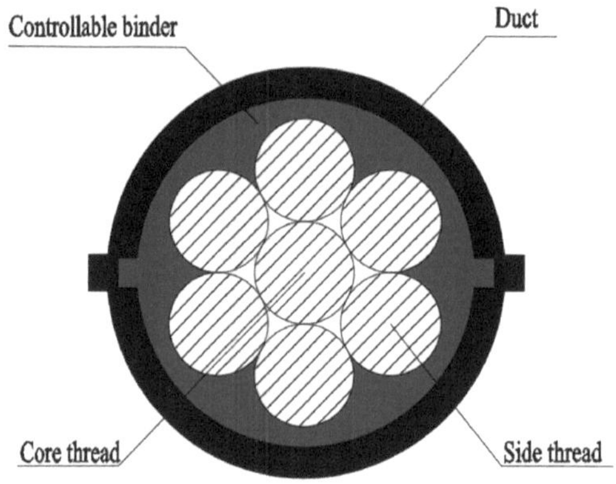


**Figure 1.** Profile construction of controllable bonded strands.

The steel strand stands as a notable conductor of electricity and heat. Leveraging the principle of prestressed reinforcement's thermal expansion and contraction, the electrothermal method initially saw application in applying mild prestress to early stage pre-tensioned steel strands, and it has now become a staple for structural reinforcement purposes [7]. However, practical engineering usage reveals considerable discrepancies in temperature field distribution and elongation value calculations for prestressed reinforcements, stemming from the electrothermal method. This, combined with the absence of a robust theoretical foundation, significantly curtails the widespread adoption and utility of the electrothermal method within prestressed tensioning approaches.

Epoxy resin stands as a prevalent bonding agent, widely employed to enhance adhesion between the steel strands and duct, thereby bolstering the bond strength [8–10]. The efficacy of epoxy adhesives hinges upon the application technique and environmental conditions [11,12]. Temperature significantly influences the properties of controllable binders, given their pronounced temperature-dependent behaviors [13]. Notably, epoxy as an individual material exhibits accelerated curing at higher temperatures [14,15]. Research reveals that curing durations needed to achieve ultimate resin strength are halved with every 10 °C increase in curing temperature beyond 5 °C [16].

Curing temperature profoundly impacts both the stiffness and strength progression rates, yielding a non-monotonic effect on ultimate mechanical properties. While raising the curing temperature to the glass transition temperature ($T_g$) augments strength and stiffness, surpassing $T_g$ triggers degradation due to heightened crosslinking network randomness, oxidative crosslinking, and polymer structure deterioration [17–19]. Thus, effective control of the controllable binder's temperature can curtail the curing time and amplify adhesive strength.

In light of the heightened temperature sensitivity exhibited by controllable adhesive materials and the practical viability of electrothermal processes, this paper takes a pioneering approach by conducting electrothermal testing on controllable adhesive strands. The primary objective is to apply controllable temperature stimulation to prestressed strands through electric heating. This method capitalizes on the generation of Joule heat within the prestressed strands, which is subsequently transferred to the surrounding controllable adhesive coating. The resultant increase in the internal temperature of the controllable adhesive facilitates a reduction in the curing time, thereby enabling controllable bonding. In this context, a comprehensive analysis of the temperature field distribution characteristics of controllable adhesive prestressed strands is undertaken, encompassing experimental investigation, simulation studies, and theoretical analyses. This multifaceted exploration aims to unravel the intricate mechanisms underlying controllable bonding achieved through the innovative integration of electrothermal techniques.

Through experimentation and simulation analysis, Cai et al. [20] introduced a method to accurately forecast the evolution of concrete hydration heat temperature fields during steam curing. Jiang et al. [21] simulated the temperature fields of prestressed concrete components subjected to conductive heating, suggesting the potential application of infrared detection for identifying quality issues in prestressed concrete voids. Wei et al. [22] conducted numerical simulations and field measurements to validate the temperature fields of a continuous concrete box girder bridge, showcasing excellent agreement between calculated and measured values. This collective body of work underscores the efficacy of a numerical simulation and field measurement verification in comprehensively analyzing temperature field distributions within prestressed structures.

Consequently, this study undertakes practical measurements of electrically heated temperatures in controllable bonded prestressed reinforcement, accompanied by numerical simulations of the temperature field. A comparative analysis of method feasibility is performed. Furthermore, addressing the dearth of theoretical research on the electrothermal method, a theoretical solution and calculation approach is proffered, grounded in heat transfer principles. This contribution establishes a foundation for future advancement and application of the electrothermal method for controllable bonded reinforcement.

## 2. Temperature Testing of Controllable Bonded Prestressed Members

### 2.1. Test Profile

The study is based on actual engineering requirements, focusing on a large, prefabricated bus station primarily utilized for bus parking and maintenance. It is characterized by a substantial load and extensive beam spans. Initially, a portion of the strands undergo pre-tensioning to counteract the self-weight effects in order to elevate the beam to its designated position. Subsequently, the remaining strands are post-tensioned to enhance its load-bearing capacity. To significantly reduce construction duration and improve project quality, the electrothermal method is being considered to expedite binder curing time, enabling controllable tensioning.

To verify the feasibility of this method, a controllable bonded prestressed beam with a size of 400 × 900 × 14,400 mm was chosen for the experiment. There were eight controllable bonded prestressed strands in the beam, with a diameter of 15.2 mm, a strength level of 1860 MPa, and a nominal sectional area of 140 mm$^2$. The upper part of the beam was outfitted with 4 Φ 25 longitudinal reinforcements, while the lower segment featured 7 Φ 25 longitudinal reinforcements and Φ 8 stirrups. In addition, there is a bundle of four 15.2 mm pre-tensioned prestressed steel strands before hoisting with a strength level of 1860 MPa.

Within the tested structure, all prestressed strands in the beam exhibit a strength of $f_{ptk}$ = 1860 MPa. Given the structure's prefabricated nature, a controlled tensioning procedure is employed before hoisting the test beam. This process involves tensioning the four lowermost prestressed strands to achieve a precise stress level of 0.75 fptk, effectively countering the self-weight effects of the beam. Figure 2 illustrates the layout of the experimental beam's reinforcement.

Furthermore, the controllable bonded prestressed strands are embedded within the concrete before the concrete is poured. Tensioning of these strands is initiated 28 days after the concrete is poured, coinciding with the concrete's achievement of a strength of 43 MPa from the same batch. The tensioning of the controllable bonded prestressed strands is closely regulated to maintain a control force of 0.75 $f_{ptk}$. The tensioning process follows a systematic sequence of 20% of $f_{ptk}$, 50% of $f_{ptk}$, and ultimately 100% of $f_{ptk}$, ensuring the precision and quality of the tensioning process. This comprehensive approach includes the mutual verification of tension force and elongation values, thereby upholding the standards of accuracy and reliability.

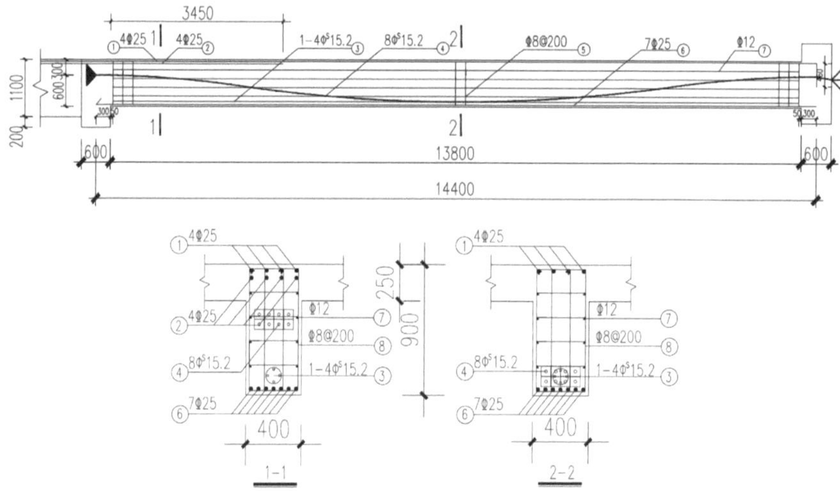

**Figure 2.** Experimental beam reinforcement layout.

## 2.2. Experimental Setup

The steel-strand power supply operates in a constant current mode, facilitating precise control over the heating power. The electric heating apparatus adopts the ZFJDR30500T (Beijing Zhaofuji New Material Technology Development Co., Ltd., Beijing, China) controllable bonded electric heater, which provides real-time output of temperature load data such as current and voltage. Each controllable bonded strand is paired with an XMZ-JK8 (Zhejiang Yitai Temperature Control Instrument Factory, Yuyao, China) eight-channel thermometer. Along the prestressed strand, eight measurement points are strategically positioned, each hosting a Pt-100 (Zhejiang Yitai Temperature Control Instrument Factory) temperature sensor. Additionally, a BHB120-3CA250 (Jinan Technology Co., Ltd., Jinan, China) high-temperature resistance strain gauge is positioned axially along the core wire. For the purpose of facilitating data analysis, sensor names have been defined based on their respective locations. One sensor positioned at the tensioning end is designated as Sensor 8, while another located at the anchorage end is referred to as Sensor 1. These two sensors are collectively termed 'terminal point sensors'. Furthermore, a sensor placed at the contact point between the anchorage end steel strand and the concrete is identified as Sensor 2 and labeled as the 'contact point sensors'. The remaining five sensors are equally spaced along the steel strand inside the beam, dividing the beam length into six equal segments. Among these, Sensor 5 is situated in the middle, denoted as the 'midpoint sensors', while the remaining sensors (3, 4, 6, and 7) are designated as 'six equinoctial point sensors', signifying the six equal segments. Figure 3 represents the arrangement of heating points, while Figure 4 displays the actual sensor placement.

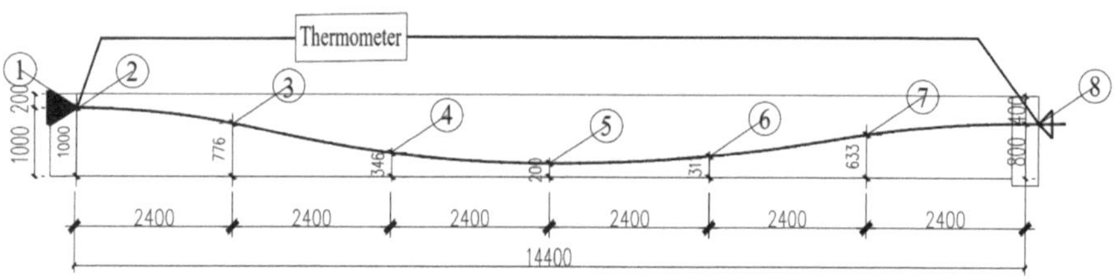


**Figure 3.** Heating point configuration.

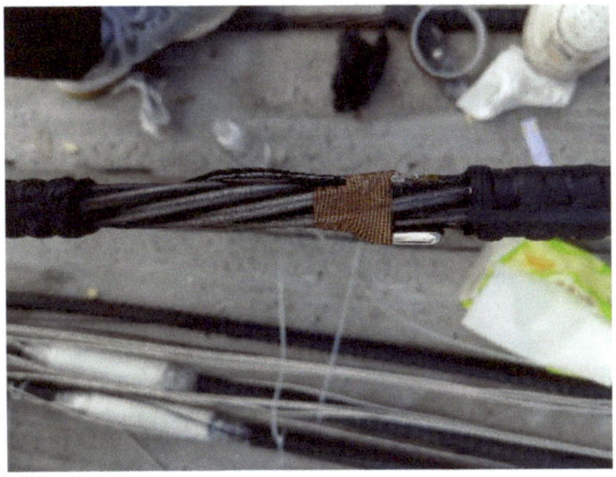

**Figure 4.** Sensor placement.

### 2.3. *Operating Conditions*

While heating the controllable bonded strand, ensuring minimal degradation of internal components, particularly the steel strands, at the target temperature is imperative. Existing research indicates that concrete's thermal performance remains relatively stable over rising temperatures [23,24]. Notably, steel strands exhibit a consistent trend of decreasing ultimate strength as the temperature rises [25,26]. At 200 °C, ultimate strength, yield strength, and elastic modulus of steel strands experience slight reductions—only 3.6%, 7.6%, and 6.0%, respectively [27–29]. Consequently, it is established that the maximum permissible temperature for steel strands during heating must not exceed 200 °C.

Serving as the essential element within prestressed concrete, the binder experiences a decline in viscosity with rising temperatures. Heating contributes to a reduction in the binder viscosity to a certain extent. The latent heat-sensitive curing agent blended within necessitates an internal temperature of at least 60 °C for interaction with other binder components. As a result, it is stipulated that the internal temperature of binder-bonded prestressed concrete must not fall below 60 °C during the heating process.

Upon establishing the heating target temperature, the heating process necessitates a careful assessment of both heating efficiency and duration. The insufficient current might delay reaching the target temperature, while an excessive heating time could prove impractical for engineering applications. Conversely, an excessive current can result in overly rapid heating, leading to temperature limits being surpassed swiftly, causing an uneven steel-strand temperature distribution. Similarly, an overly brief heating time hampers timely adjustments based on real-time conditions. Striking the right balance is crucial for effective heating. Following advanced modeling analysis and theoretical calculations, the operating conditions are as listed in Table 1. Figure 5 shows how the electrothermal method is applied to prestressed strands.

**Table 1.** Setting of operating conditions.

| Target Temperature | Current Magnitude | ID |
|---|---|---|
| 200 | 350 | 200-350-a |
| 200 | 350 | 200-350-b |
| 200 | 300 | 200-300-a |
| 200 | 300 | 200-300-b |
| 150 | 350 | 150-350-a |
| 150 | 350 | 150-350-b |
| 150 | 300 | 150-300-a |
| 150 | 300 | 150-300-b |

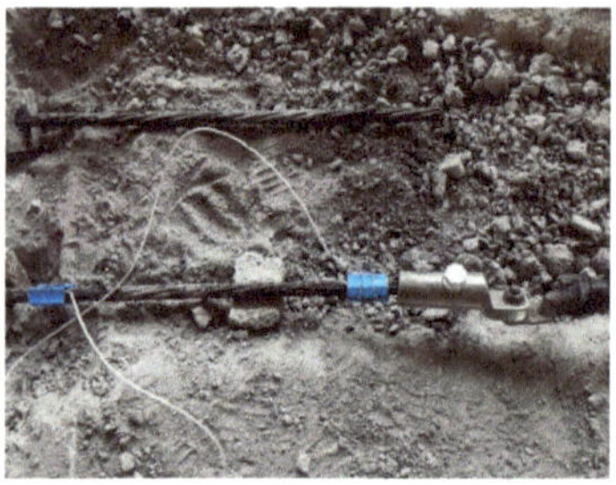

**Figure 5.** Electrically heated controllable bonded strand.

Consequently, the operational parameters are outlined in Table 1, featuring heating currents of 300 A and 350 A, with corresponding target temperatures set at 150 °C and 200 °C. Given the inherent uncertainty in the temperature field distribution during heating, repeated experiments are imperative to ensure the dependability of the obtained results.

According to the target temperature and current magnitude, each operating condition is named in the form of target temperature-current magnitude. The suffix 'a' is attached to the first test under the same conditions, and the repeated test is 'b'. Specific operating conditions are shown in Table 1.

During the experiment, the heating end is directly exposed to air through convection, leading to faster heating compared to the prestressed reinforcement's temperature within the concrete. Consequently, reaching the temperature limit becomes more accessible. To streamline measurements and progress determination in prospective engineering applications, the temperature at the exposed concrete end is chosen as the control parameter. Crucial temperature-related parameters requiring control are systematically gathered and uploaded by the temperature controller. Additionally, the real-time output of heating current, voltage, and duration is facilitated by the controllable bonded heating equipment.

## 3. Test Results and Discussions

To mitigate the impact of forced convection heat transfer from air currents on the exposed concrete at the end of controllable bonded prestressed tendons, a layer of nano-aerogel felt was applied to the exterior of the end steel strand during the heating process as depicted in Figure 6.

**Figure 6.** Insulated strand termination.

The temperature controller displays and records the real-time temperature of each point throughout the heating process, depicted in Figure 7. As heating ensues, there is a swift rise in temperature at the end of the controllable bonded strand, while the controllable bonded strand embedded within the concrete experiences a comparatively gradual temperature increase.

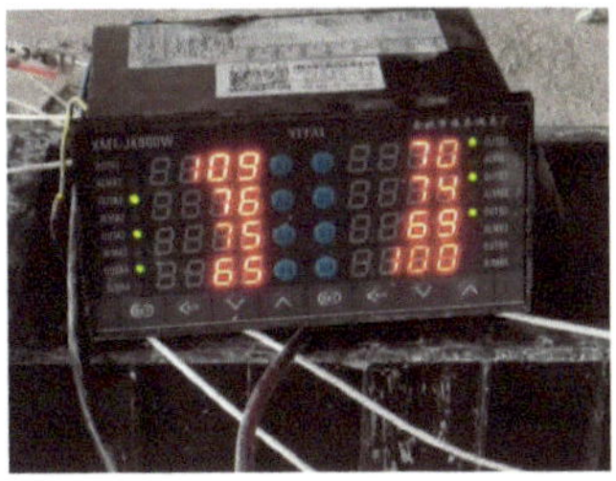

**Figure 7.** Thermometer measured temperature.

Utilizing the temperature sensor test data, thermal diagrams of prestressed tendons are generated for current sizes of 350 A and 300 A, as illustrated in Figures 8 and 9, respectively.

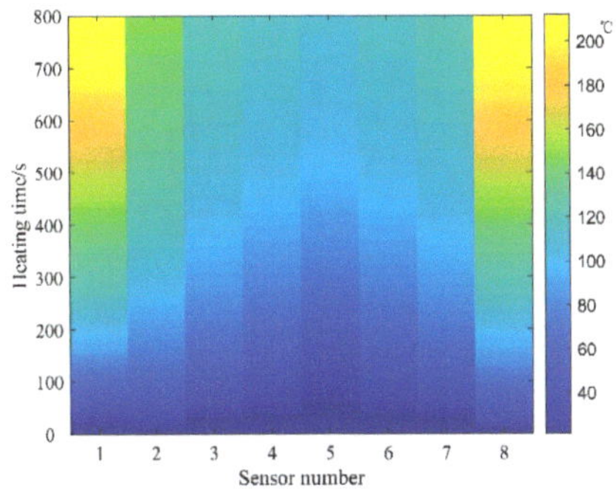


**Figure 8.** Temperature distribution at 350 A.

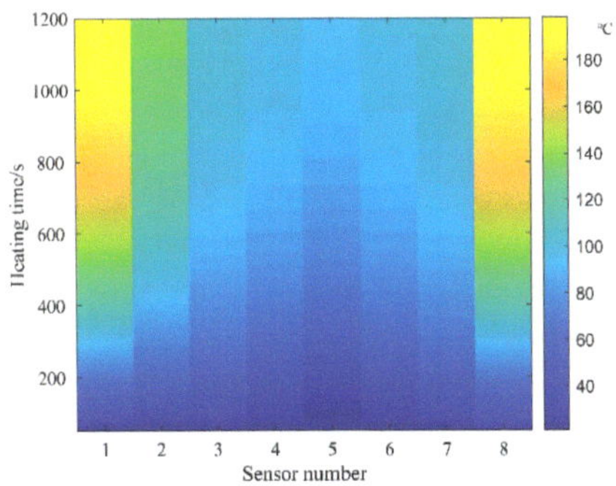


**Figure 9.** Temperature distribution at 300 A.

The thermal diagrams reveal that during the actual measurement process, when excluding the impact of inadvertently forced air convection heat transfer, the temperature distribution of prestressed ribs demonstrates a notable symmetrical pattern. Temporally, after activation, temperatures at both ends experience a swift escalation, progressively augmenting the temperature difference between the ends and the mid-span of the prestressed ribs. Spatially, within the DC heating process, temperatures of the prestressed ribs from both ends to the mid-span experience a slight decline, while the temperature distribution remains uniform within the concrete. This behavior arises due to the curved symmetrical arrangement of the prestressed ribs, with proximity to the mid-span resulting in a reduced distance between the rib surface and concrete surface, consequently expediting heat dissipation.

Regarding heating efficiency, despite the current increase from 300 A to 350 A representing a mere 16.6% rise, the heating time diminishes from 1200 s to 800 s—a reduction of 50%. This underscores the pivotal role of heating current in determining efficiency. Thus, an imminent necessity arises for formulating an electric heating strategy for controllable bonded prestressed ribs.

Given the near-symmetrical temperature field distribution of the prestressed strands, the analysis focuses on select measuring points—namely the endpoint, contact point, sixth point, and midpoint—located on one side. Figure 10 illustrates the temperature–time curves of the prestressed strands subjected to the electrothermal method under varying current sizes.

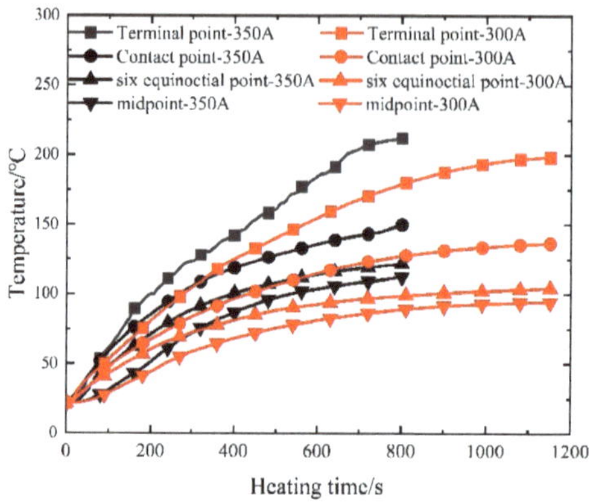


**Figure 10.** Measured temperature–time curve of prestressed steel strands.

As depicted in Figure 10, the initial 400 s of heating exhibit an accelerated temperature rise in the prestressed reinforcement, proportionate to the current increase. Subsequent heating showcases nearly identical temperature ascent rates for prestressed reinforcement within concrete, regardless of current size, ultimately converging to a stable temperature. This observation underscores that varying current sizes wield a more pronounced impact during the early stages of electric heating, exerting a diminishing influence in later stages.

At the designated measuring points within the concrete's interior, high-temperature strain gauges were positioned. Figure 11 presents the strain–time curves of prestressed ribs under diverse current sizes.

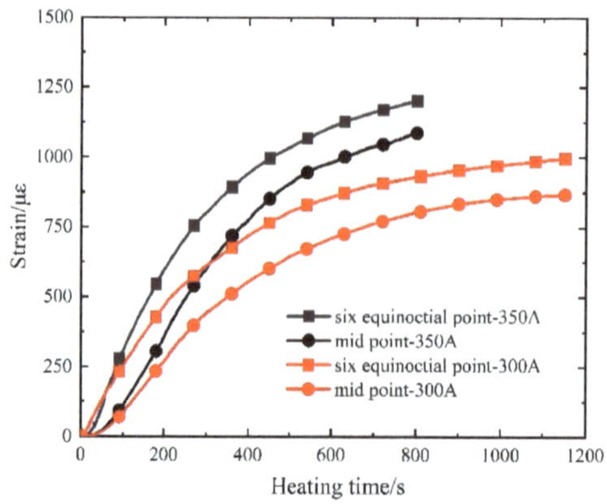


**Figure 11.** Measured strain–time curve of prestressed steel strands.

While the controllable bonded binder remains in a non-solidified state during heating, the prestressed steel strand retains the capacity for unimpeded movement within the binder. The temperature-induced strain of the prestressed steel strand can be computed using Equation (1), where the thermal expansion coefficient ($\alpha$) of the prestressed steel strand stands at $1.2 \times 10^{-5}/°C^{-1}$.

$$\varepsilon = \alpha \cdot \Delta T \tag{1}$$

As shown in Table 2, the calculated maximum relative error stands at 9.10%, observed at the sixth point under the 350 A operating condition. For all other measuring points, the maximum error is 0.83%, demonstrating a high level of reliability in the test's temperature measurements.

**Table 2.** Temperature strain comparison.

| Measuring Point | ΔT/°C | Theoretical Strain Value/με | Measured Strain Value/με | Relative Error/% |
|---|---|---|---|---|
| Six equinoctial points-350 A | 110.2 | 1322.4 | 1202.01 | −9.10 |
| Mid-point-350 A | 90.6 | 1087.2 | 1087.68 | 0.04 |
| Six equinoctial points-300 A | 82.3 | 987.6 | 995.83 | 0.83 |
| Mid-point-300 A | 72.4 | 868.8 | 866.136 | −0.31 |

A comprehensive analysis of the obtained results reveals the following under a continuous connection to a constant current during heating:

1. The temperature field within prestressed concrete showcases symmetry, yielding a more uniform temperature distribution. Despite the concrete's poor thermal conductivity, measuring points nearer the concrete surface exhibit lower temperatures, indicative of higher heat loss.
2. Larger currents lead to swifter temperature elevation in controllable bonded prestressed reinforcement upon power activation. However, with smaller currents, the heating power deficiency intensifies the heat transfer between the controllable bonded prestressed reinforcement and its surroundings as the temperature rises. In the later stages, the prestressed reinforcement attains a balance between the temperature rise and heat dissipation, resulting in a quicker approach to stable temperature conditions. During this phase, despite ongoing electric heating, the temperature remains nearly constant at a certain level, fostering favorable conditions for the curing reaction of the controllable bonded binder within this environment. The comparison of stable temperatures under different current sizes is tabulated in Table 3. By evaluating the ratio of stable temperature to current size, a positive correlation between the stable temperature and current size becomes apparent. Notably, changes in current size exert a significant influence on the endpoint temperature while impacting the prestressed reinforcement within the concrete to a relatively lesser extent.
3. The temperature variation trend within the exposed concrete of controllable adhesive tendons mirrors that of the interior, albeit with higher actual temperatures. In practical applications where internal temperature sensing is unfeasible, estimating internal temperature can be approximated by measuring the contact point between prestressed tendons and concrete.
4. The measured temperature-induced stress aligns remarkably well with theoretical values, indicating that during electric heating of prestressed strands, steel strands undergo uninhibited elongation within the controllable bonded binder. This observation implies that temperature-induced stress can be leveraged to impart lower prestress, thus fostering a more even distribution of effective prestress throughout the prestressed strands.

**Table 3.** Stable temperature comparison.

| Current Magnitude | Measurement Point | Stable Temperature/°C | Temperature/Current |
|---|---|---|---|
| 350 A | Terminal point | 212.2 | 0.6057 |
| 350 A | Contact point | 150.0 | 0.4286 |
| 350 A | Six equinoctial point | 121.9 | 0.3483 |
| 350 A | Midpoint | 112.3 | 0.3286 |
| 300 A | Terminal point | 198.3 | 0.6610 |
| 300 A | Contact point | 136.6 | 0.4553 |
| 300 A | Six equinoctial point | 104.0 | 0.3366 |
| 300 A | Midpoint | 94.2 | 0.3140 |

## 4. Temperature Field Simulation and Analysis

### *4.1. Model Development*

#### 4.1.1. Model Overview and Description

FLUENT, a comprehensive fluid dynamics computation software, is harnessed for simulating intricate fluid and heat transfer phenomena. In this study, FLUENT is employed to conduct numerical simulations of the temperature field distribution within prestressed reinforcement during the electrothermal method. This endeavor aims to delve into the governing principles and influencing factors governing temperature distribution in a prestressed reinforcement.

Utilizing the test results and disregarding the impact of thermal conductivity from regular steel strands within the beam, an analysis is conducted on a controllable bonded prestressed beam measuring 400 × 900 × 14,400 mm. The beam accommodates eight controllable bonded prestressed strands, each possessing a diameter of 15.2 mm, a strength level of 1860 MPa, and a nominal cross-sectional area of 140 mm$^2$. The material parameters pertinent to the model are detailed in Table 4.

**Table 4.** Material parameters for the numerical model.

| Materials | Thermal Conductivity/(W/(m·K)) | Specific Heat/(J/(kg·K)) | Density/(kg/m$^3$) |
|---|---|---|---|
| Steel strand | 49 | 470 | 7800 |
| Binder | 0.2 | 1600 | 1300 |
| Duct | 0.5 | 2300 | 980 |
| Concrete | 1.74 | 920 | 2500 |

#### 4.1.2. Mesh Generation

During the process of grid division, accommodating the spiral nature of the steel strand and achieving a controllable bonded binder's coating pose challenges for high-order curved surfaces. The attainment of high-precision grid calculations that easily converge is intricate. To address this issue, an approximation is made by treating the cross-section of the steel strand as a circle with a diameter matching its nominal diameter. Similarly, the cross-sections of the controllable bonded binder and duct are considered as concentric rings with specific thicknesses. Due to significant differences in size between external concrete and prestressed reinforcement, model division involves grids with considerably varied aspect ratios, resulting in a substantial number of grids.

To navigate these complexities, the model division predominantly employs a hexahedral grid, with an emphasis on denser grid allocation for intricate details. The numerical representation of electric heating for the controllable bonded binder is depicted in Figure 12, and the grid partitioning is illustrated in Figure 13.

**Figure 12.** Electrothermal model for controllable bonded reinforcement.

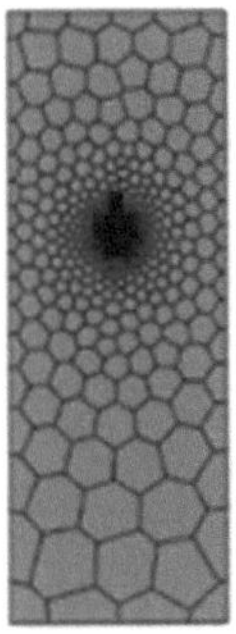

**Figure 13.** Segmented meshing for an electrothermal model of controllable bonded strands.

Due to the significantly low thermal conductivity of the sheathing material utilized for controllable bonded prestressing strands, the influence on other prestressing strands is negligible when heating a single one. In order to optimize the efficiency of the numerical model and streamline subsequent theoretical analysis, each individual prestressing strand is independently examined within the model.

#### 4.1.3. Applied Loads and Boundary Conditions

During the application of model loads, one end of the steel strand is designated as a voltage zero potential surface, while the other end receives a current load. Simultaneously, temperature continuity is ensured at the interface of each material.

For establishing boundary conditions, natural convection heat transfer occurs among concrete, steel strand, and air at the heat transfer boundary. This falls within the third category of thermal analysis boundary conditions. The convection heat transfer coefficient between the concrete surface and air is determined as 4.74 $W/(m^2{\cdot}K)$ [30].

The typical formula for calculating the air convection heat transfer coefficient [31] can be expressed as

$$h = N_u \times k/L \tag{2}$$

where $h$ represents the air convection heat transfer coefficient in $W/(m^2{\cdot}K)$, and $N_u$ signifies the dimensionless Nusselt number—an inclusive indicator of fluid flow state and heat transfer characteristics. Additionally, $k$ denotes the thermal conductivity of air in $W/(m{\cdot}K)$, and $L$ signifies the characteristic length, generally referring to the characteristic length of the heat-exchanging object's surface in meters.

Taking into account that the duct and the controllable bonded binder have limited heat transfer areas on both ends of the concrete, and due to uniform heat transfer, the surface can be approximated as smooth, thereby yielding a constant air convection heat transfer coefficient. Accordingly, the calculated convection heat transfer coefficient between the controllable bonded binder and air is 17 W/($m^2$·K), and the calculated convection heat transfer coefficient between the duct and air is 21 W/($m^2$·K).

The energy equation is engaged during computation, incorporating thermoelectric coupling calculations. The initial model temperature is set at 22 °C. For optimizing energy consumption and efficiency, based on test results, the simulation calculation duration is established as 800 s for a current of 350 A, and 1200 s for a current of 300 A. The time step is designated as 1 s. Notably, the calculation overlooks the influence of temperature-induced deformation of the steel strand on concrete stress.

### 4.2. Simulation Results and Experimental Validation

In the test, the steel strand's span diameter is significant, leading to its division into two distinct sections during the heating process: the portion enveloped by concrete and the exposed section. The exposed part is notably shorter, resulting in discernible temperature distribution differences between the two segments. The primary research emphasis centers on the controllable bonded binder segment within the concrete, particularly focusing on the temperature field of the steel strand and the controllable bonded binder.

As an illustration, let us consider operating condition 200-350-a. Figure 14 presents the post-heating temperature field distribution across the entire beam. Notably, both the measured and simulated results exhibit symmetrical distributions. To validate this, simulated values from one side's measuring point are compared with the corresponding measured data.

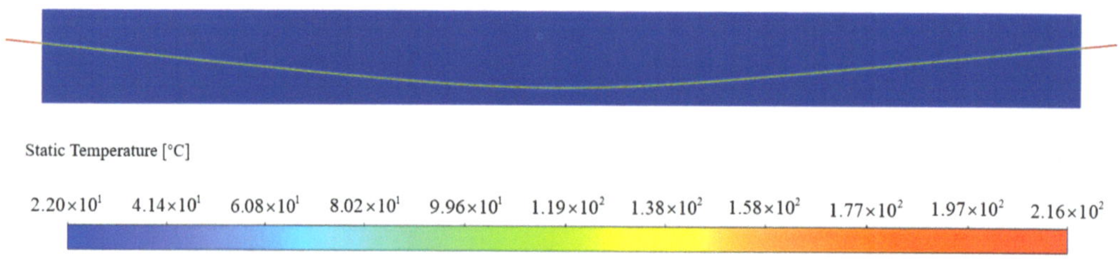


**Figure 14.** Numerical simulation of temperature field distribution in whole beam.

From Figure 14, several key observations can be made regarding the temperature distribution during heating:

The outer concrete showcases uniform temperature distribution during the heating process, approximating room temperature.

(1) The steel strand's endpoint experiences high temperatures, with the maximum temperature reaching 216 °C.
(2) The temperature decreases rapidly along the steel strand from its endpoint to the contact point with the concrete. At the contact point, the temperature registers at 150 °C.
(3) The temperature distribution within the steel strand inside the concrete manifests symmetry. Notably, the temperature slightly diminishes towards the middle of the span. For instance, the sixth point reflects a temperature of 121.9 °C, while the lowest temperature of 112.3 °C occurs at the midpoint.

The temperature at the interface between the steel strand and the controllable bonded binder remains consistently continuous. When a constant current (350 A, 300 A) is continuously applied to the steel strand, the temporal evolution of temperature at each measuring point is illustrated in Figure 15.

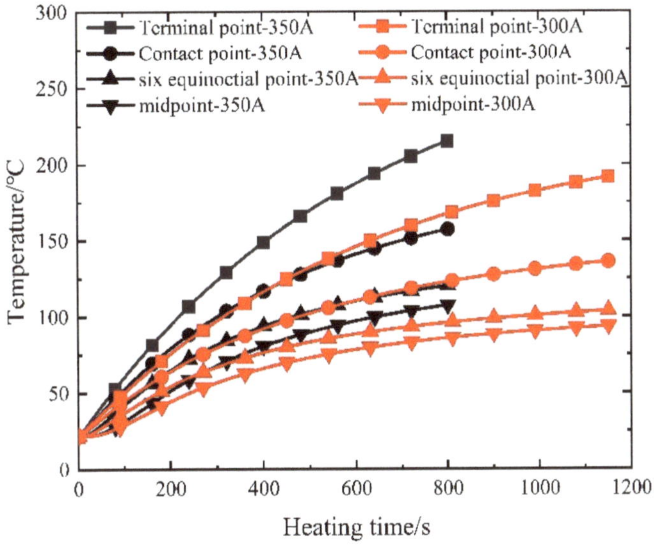


**Figure 15.** Simulated temperature–time curve.

The simulation outcomes closely mirror the measured results. Notably, the current size wields a substantial impact on the initial heating rate, which progressively diminishes as time elapses. As the temperature of the controllable bonded prestressed reinforcement escalates, heat exchange with the surrounding environment intensifies. Consequently, the prestressed reinforcement attains a heating–radiative balance more swiftly, resulting in temperature stabilization within a plateau phase.

Figures 16 and 17 depict the juxtaposition between the measured and simulated temperature field values across varying current sizes.

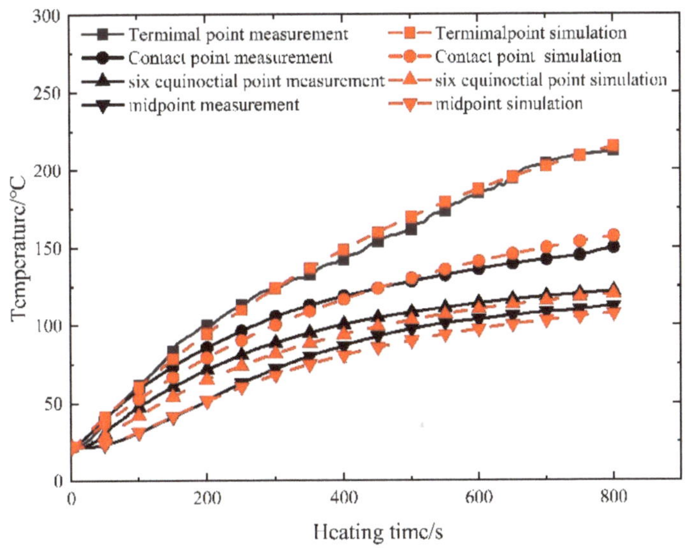


**Figure 16.** Comparison of measured and simulated temperature for prestressed strands at 350 A.

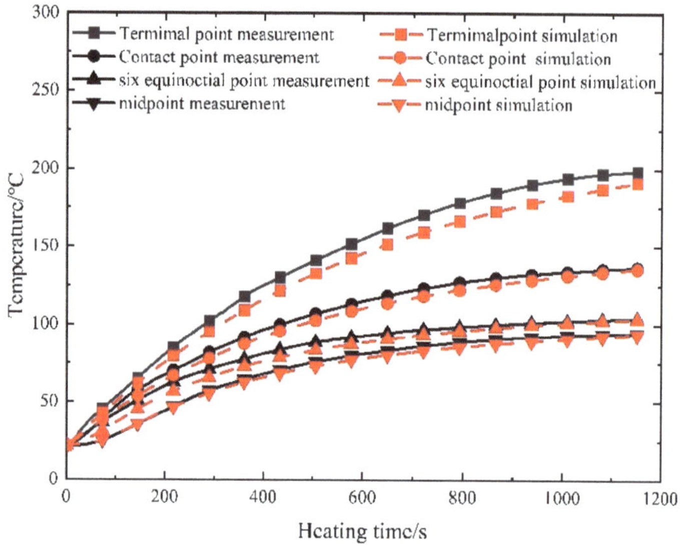


**Figure 17.** Comparison of measured and simulated temperature for prestressed strands at 300 A.

The root-mean-square error (RMSE) and mean absolute error (MAE) for both the measured and simulated curves at each measuring point across different current sizes have been calculated and are presented in Table 5.

**Table 5.** Comparison of measured and simulated temperature fields.

| Current | Measurement Point | Root Mean Square Error/°C | Average Absolute Error/°C | Maximum Relative Error |
|---|---|---|---|---|
| 350 A | Terminal point | 4.11 | 3.54 | 1.38% |
| 350 A | Contact point | 4.29 | 3.72 | 0.91% |
| 350 A | Six equinoctial point | 4.19 | 3.79 | 1.39% |
| 350 A | Midpoint | 3.97 | 3.18 | 1.18% |
| 300 A | Terminal point | 8.89 | 8.37 | 1.73% |
| 300 A | Contact point | 3.71 | 3.51 | 0.68% |
| 300 A | Six equinoctial point | 4.00 | 3.59 | 1.19% |
| 300 A | Midpoint | 2.17 | 1.90 | 0.82% |

The temperature deviation between the measured and simulated temperatures at the terminal point is relatively significant. This discrepancy primarily arises from the direct exposure of the terminal point to the surrounding air, where airflow conditions directly impact the convective heat transfer coefficient, thereby significantly influencing endpoint temperatures. Moreover, when setting the heating current to 300 A, lower heating power results in a reduced capacity to counteract variations in heat conduction, leading to a larger temperature difference. Nevertheless, despite these factors, the relative error remains relatively small. In contrast, internal steel strands within the concrete exhibit a closer alignment between measured and simulated values, as they are insulated from direct air contact.

Apart from the terminal points potentially experiencing considerable influence from air flow heat transfer, the maximum root-mean-square error and mean absolute error for other measuring points amount to 4.29 °C. A comparison of the curves further underscores that the model's maximum relative error is 1.73%, with the error diminishing over prolonged time intervals—indicating the heightened precision of the simulation method. The test results affirm the commendable reliability of the numerical model for controllable bonded electric heating. Particularly, for lower currents, an increase in heating time corresponds to a more stabilized temperature rise and heat transfer within the prestressed tendons, with the measured outcomes aligning closely with simulation results.

The improvement in simulation accuracy relies on the refinement of grid partitioning. In this numerical model, local mesh refinement has been applied to the controllable binder regions, and convective heat transfer coefficients have been adjusted based on measured data to enhance the accuracy. However, certain factors affecting precision were not accounted for in the simulation, including interfacial thermal resistance between material layers, contact resistance at the heating terminals, and potential incomplete insulation of the pads, which might lead to a reduction in accuracy.

### *4.3. Parameter Analysis*

For a comprehensive grasp of the temperature field distribution and application approach of the controllable bonded prestressed beam electrothermal method, electric loads are administered at both ends of the prestressed beam. Subsequently, a verified finite element model is employed for thorough parameter analysis. The scrutinized and assessed parameters encompass the current size, thickness of the controllable bonded binder, steel strand diameter, and other pertinent factors.

#### 4.3.1. Current Size

When adjusting the current density applied to the steel strand end face, the curing reaction occurs solely upon reaching the specified internal temperature within the controllable bonded binder. This approach entails a controllable temperature elevation. Given an initial model temperature of 20 °C, the heating current varies between 200 A and 500 A in increments of 50 A to determine the requisite heating time. The simulation outcomes are then employed to construct a scatter diagram. The data are fitted using a power function in the form of $y = a{\cdot}x^b$. The fitting results are showcased in Figure 18, with parameters $a = 8.9784 \times 10^9$ and $b = -2.8891$.

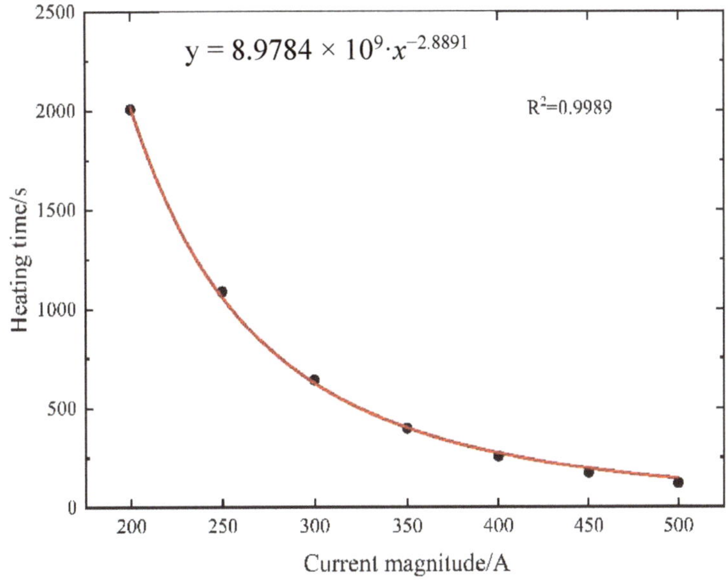


**Figure 18.** Current–time curve for midpoint temperature rise to 80 °C in prestressed strands.

The value of $b = -2.8891$ suggests that while the Joule heat generated by the steel strand heating method exhibits proportionality to the current, the heat transfer characteristics of the material comprising the prestressed strands amplify with their temperature ascent. Consequently, when subjected to a small current, the prestressed strands tend to achieve thermal equilibrium more readily during heating. Conversely, employing a larger current yields greater heating power and facilitates easier attainment of the target temperature.

Nevertheless, it is important to exercise caution when increasing the current solely to enhance the heating efficiency. The curing reaction does not occur instantaneously; it typically necessitates maintaining specific temperature conditions for 5–10 min. Overreliance on high currents could lead to prolonged heating of the prestressed strands, causing the final temperature to become excessively elevated. This, in turn, might compromise material performance and hinder the intended curing effect. Hence, a balanced approach is crucial to achieve effective heating while ensuring optimal material behavior and desired curing outcomes.

#### 4.3.2. Controllable Bonded Binder Thickness

The controllable bonded binder serves as the core element. An excessively small thickness might lead to an inadequate bond with the prestressed steel strands. Conversely, an overly large thickness could prolong the curing time and result in material wastage. Currently, the binder thickness is distributed between 2 mm and 3.5 mm. Accordingly, the simulation employs a step size of 0.25 mm and heats various controllable bonded binder thicknesses for equivalent durations and currents.

As shown in Figure 19, the controllable bonded binder exhibits a notable behavior as its thickness is altered. Due to its low thermal conductivity, high thermal resistance, and effective thermal insulation properties, augmenting the binder's thickness yields a roughly linear increase in its thermal resistance within a confined range. With a heightened thermal resistance, the Joule thermal conduction resistance escalates, leading to heat predominantly accumulating within the steel strand and the controllable bonded binder layer. Consequently, a temperature variance between the controllable bonded binder layer and the duct layer emerges, while the temperature of the duct layer and the concrete layer remains relatively stable.

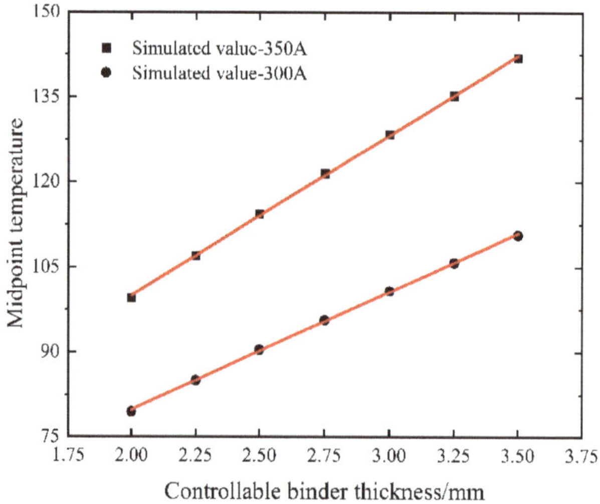


**Figure 19.** Midpoint temperature variation with controllable bonded binder thickness.

As the controllable bonded binder layer thickness increases, the midpoint temperature of the prestressed steel strand approaches the theoretical value under adiabatic conditions. Notably, the slope indicates that increasing the controllable bonded binder thickness is conducive to enhancing heating efficiency, especially when larger currents are applied. Consequently, maintaining a constant controllable bonded binder thickness and appropriately elevating the current size can effectively bolster the heating efficiency.

#### 4.3.3. Diameter of Steel Strand

The diameter of the steel strand serves as a conducting medium, exerting a direct influence on Joule heat generation, subsequently impacting the heat absorption and heating within the controllable bonded binder while keeping material properties constant. Taking into account commonly employed steel strand diameters, an electric heating simulation is conducted for prestressed strands with diameters of 17.8 mm and 21.8 mm, building upon the existing simulation involving a 15.2 mm steel strand. The results of this simulation are depicted in Figure 20.

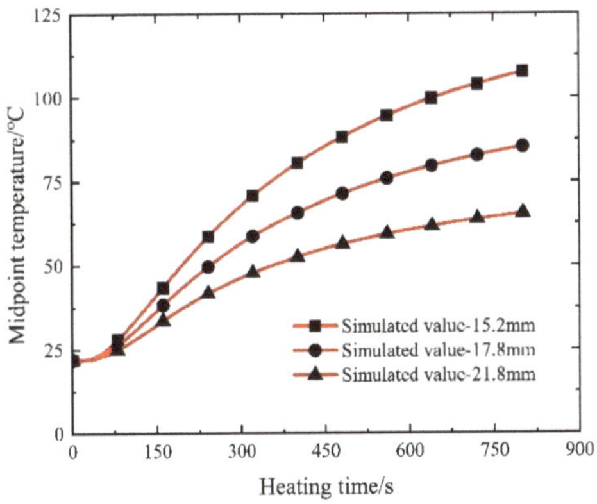


**Figure 20.** Temperature–time variation in electrically heated steel strands of varying diameters.

The presented results indicate a discernible trend: as the steel strand radius increases, the heating efficiency of the electrothermal method gradually diminishes, and larger-diameter controllable adhesive strands reach the heating platform stage at an accelerated pace. This outcome is attributable to the unaltered resistivity and length of the steel strand: the resistance of the prestressed steel strand is inversely proportional to the square of its radius. Consequently, a larger radius results in diminished resistance and reduced thermal power.

Moreover, although the steel strand exhibits excellent thermal conductivity, an increased radius proportionally raises its thermal resistance. This amplifies the challenge of heat conduction, resulting in a decreased temperature difference between the two ends of the material and a slower heating rate. Therefore, when utilizing electric heating for controllable bonded strands with larger diameters, it becomes crucial to adjust the current magnitude accordingly through calculated means.

## 5. Theoretical Temperature Field Calculation for Controllable Bonded Prestressed Members

Previously, the viability of achieving controllable curing of controllable bonded prestressed reinforcement through electric heating was verified via field measurements. Concurrently, an accurate and effective numerical model for controllable bonded electric heating was established, yielding precise simulation outcomes. Parameter analysis elucidated the impact of the heating current, controllable bonded binder thickness, and steel-strand diameter on the electric heating approach.

However, a notable gap remains in terms of an effective calculation methodology for controlling the bonding process of controllable bonded reinforcement. To streamline the technological process of the controllable bonded prestressed reinforcement electrothermal method and to tailor the heating strategy to practical engineering contexts, the temperature field distribution of controllable bonded prestressed components is theoretically computed using the Stehfest numerical algorithm.

### 5.1. Stehfest Numerical Algorithm

Stehfest introduced a formula for numerically inverting the Laplace transform [32]. Let the Laplace transform of the function $f(t)$ be denoted as $f^{-}(s)$, then the formula is given by

$$f^{-}(s) = \int_{0}^{\infty} e^{-st} f(t) dt \tag{3}$$

If $f^{-}(s)$ can be computed, the value of the function $f(t)$ at $t = T$ can be determined using the following equation:

$$f(t) = \frac{\ln 2}{T} \sum_{i=1}^{N} V_i f^{-}(\frac{\ln 2}{T} i) \tag{4}$$

where

$$V_i = (-1)^{\frac{N}{2}+i} \sum_{k=\frac{i+1}{2}}^{\min(i,\frac{N}{2})} \frac{k^{\frac{N}{2}} (2k)!}{(\frac{N}{2}-k)!k!(k-1)!(i-k)!(2k-i!)} \tag{5}$$

where $t$ and $T$ represent independent variables, $N$, $i$, and $k$ are positive integers, and $V_i$ denotes the intermediate function value. In principle, a higher number of terms in the inversion formula yields greater accuracy in calculations. However, practical considerations, especially in heat transfer calculations, often lead to selecting an even number between 8 and 20 to mitigate rounding errors.

### 5.2. Mathematical Formulation of Controllable Bonded Prestressing Electrothermal Method

Controllable bonded prestressed reinforcement constitutes a composite material with a polyphase medium as shown in Figure 21. The steel strand is composed of multiple high-strength steel wires, approximated as a cylinder with its nominal diameter for analysis. Similarly, the controllable bonded material and duct are simplified as cylindrical walls with specific thicknesses. Given that analyzing the surface temperature of concrete beams is not a primary focus and considering the limited heat transfer range due to the low thermal conductivity of the controllable bonded binder and duct, as well as the relatively low power of DC heating, for analytical convenience, the concrete is also treated as an equivalent cylindrical wall of a certain thickness.

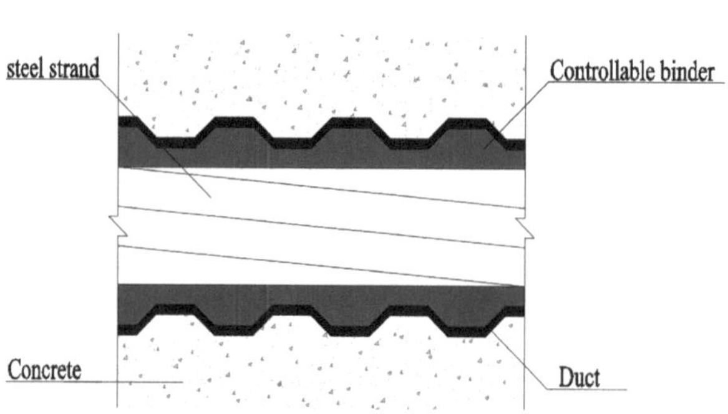


**Figure 21.** Controllable bonded prestressed tendon and concrete.

Based on heat transfer theory, the three-dimensional transient thermal conductivity temperature field $U(r, \psi, z, t)$ I, in cylindrical coordinates, is governed by

$$\rho c\frac{\partial u}{\partial t} = \frac{1}{r}\frac{\partial}{\partial r}(\lambda r\frac{\partial u}{\partial r}) + \frac{1}{r^2}\frac{\partial}{\partial \varphi}(\lambda\frac{\partial u}{\partial \varphi}) + \frac{\partial}{\partial z}(\lambda\frac{\partial u}{\partial z}) + \dot{\phi} \tag{6}$$

where $u$ is the temperature (K), $\varphi$ is the azimuthal angle (rad), $r$ is radius (m), $z$ is vertical coordinate (m), $\rho$ is density (kg/m$^3$), $c$ is specific heat (J/(kg·K)), $\lambda$ is thermal conductivity (W/(m·K)), and $\dot{\phi}$ is an internal heat source (W/m$^3$).

Considering the conventional physical properties of thermal conductivity, DC heating, and a constant, uniformly distributed internal heat source across the entire steel strand, the heat source strength follows

$$\dot{\phi} = I^2R/V \tag{7}$$

where $V$ is the volume (m$^3$), $I$ is the steady-state current value (A), and $R$ is the resistance value of the resistance ($\Omega$).

Given that the boundary conditions in the electrothermal model are angle-independent, and the length-to-diameter ratio of the steel strand in concrete ($L/d > 10$), the temperature along the steel strand is uniform along its length, neglecting longitudinal temperature changes. Consequently, the temperature field becomes a function $u(\varphi, r)$ of $\varphi$ and $r$.

Hence, the original problem of DC heating in controllable bonded prestressed strands can be simplified into a one-dimensional transient heat transfer problem of a multi-layer composite cylindrical wall. This simplification can be represented by the following heat conduction equations:

$$\begin{cases}
\dfrac{\partial u_1}{\partial t} - a_1\dfrac{1}{r}\dfrac{\partial}{\partial r}(r\dfrac{\partial u}{\partial r}) = \dfrac{I^2R}{\rho_1 c_1 V}, 0 \le r \le r_1 \\
\dfrac{\partial u_2}{\partial t} - a_2\dfrac{1}{r}\dfrac{\partial}{\partial r}(r\dfrac{\partial u}{\partial r}) = 0, r_1 \le r \le r_2 \\
\dfrac{\partial u_3}{\partial t} - a_3\dfrac{1}{r}\dfrac{\partial}{\partial r}(r\dfrac{\partial u}{\partial r}) = 0, r_2 \le r \le r_3 \\
\dfrac{\partial u_4}{\partial t} - a_4\dfrac{1}{r}\dfrac{\partial}{\partial r}(r\dfrac{\partial u}{\partial r}) = 0, r_3 \le r \le r_4 \\
\left.\dfrac{\partial u_1}{\partial t}\right|_{r=0} = 0 \\
\lambda_1\left.\dfrac{\partial u_1}{\partial r}\right|_{r=r_1} = \lambda_2\left.\dfrac{\partial u_2}{\partial r}\right|_{r=r_2} \\
u_1(r_1, t) = u_2(r_1, t), \\
\lambda_2\left.\dfrac{\partial u_2}{\partial r}\right|_{r=r_2} = \lambda_3\left.\dfrac{\partial u_3}{\partial r}\right|_{r=r_2} \\
u_2(r_2, t) = u_3(r_2, t), \\
\lambda_3\left.\dfrac{\partial u_3}{\partial r}\right|_{r=r_3} = \lambda_4\left.\dfrac{\partial u_4}{\partial r}\right|_{r=r_3} \\
u_3(r_3, t) = u_4(r_3, t), \\
-\lambda_4\left.\dfrac{\partial u_4}{\partial r}\right|_{r=r_4} = h(u_4 - u_{ext})
\end{cases} \tag{8}$$

where $r$ is the radius(m); $r_1$, $r_2$, $r_3$, and $r_4$ are the coordinates of the steel strand, the controllable adhesive material, the duct, and the outer surface of the concrete, respectively; $u_1$, $u_2$, $u_3$, and $u_4$ are the temperature of the steel strand (K), the controllable adhesive material, the duct, and the outer surface of the concrete, respectively; $a_1$, $a_2$, $a_3$, and $a_4$ are the thermal diffusivity of the steel strand (m$^2$/K), the controllable bonded binder, the duct, and the outer surface of the concrete, respectively; $\lambda_1$, $\lambda_2$, $\lambda_3$, and $\lambda_4$ are the thermal conductivity of the steel strand, (W/(m·K)), the controllable bonded binder, the duct, and the outer surface of the concrete, respectively; $u_{ext}$ is the temperature of the fluid on the outer surface of the concrete (K); $h$ is the convection heat transfer coefficient between the outer surface of the concrete and the fluid (W/(m$^2$·K)); and $t$ is the heat transfer time (s).

The model assumes that all four materials possess constant physical properties, and the initial temperature distribution is uniformly distributed. By employing appropriate transformations, the mathematical model can be solved within the spatial domain. This mathematical framework enables an analysis of the temperature distribution and behavior of controllable bonded prestressed components during the electric heating process.

$$\begin{cases} \frac{d^2U_1}{dr^2} + \frac{1}{r}\frac{dU_1}{dr} - \beta_1 U_1 = \beta_1 U_0 + \beta_1 \frac{I^2R}{\rho c V} \\ \frac{d^2U_2}{dr^2} + \frac{1}{r}\frac{dU_2}{dr} - \beta_2 U_2 = \beta_2 U_0 \\ \frac{d^2U_3}{dr^2} + \frac{1}{r}\frac{dU_3}{dr} - \beta_3 U_3 = \beta_3 U_0 \\ \frac{d^2U_4}{dr^2} + \frac{1}{r}\frac{dU_4}{dr} - \beta_4 U_4 = \beta_4 U_0 \end{cases} \tag{9}$$

where

$$\beta_{\mathrm{i}} = \frac{\rho_i c_i}{\lambda_i}\ (i = 1, 2, 3, 4) \tag{10}$$

where $U_i$ is the Laplace transform of $u_i$; $U_0$ is the initial temperature (K); $\rho_i$ is the density of the $i$-th layer of material (kg/m$^3$); and $c_i$ is the specific heat of the $i$-th material (J/(m$^3$·K)).

The above equation is the Bessel equation [33–35], and its general solution is

$$\begin{cases} U_1(r,s) = A(s)I_0(r\sqrt{s\beta_1}) + B(s)K_0(r\sqrt{s\beta_1}) + \frac{U_0}{s} + \frac{I^2R}{s\rho c V} \\ U_2(r,s) = C(s)I_0(r\sqrt{s\beta_2}) + D(s)K_0(r\sqrt{s\beta_2}) + \frac{U_0}{s} \\ U_3(r,s) = E(s)I_0(r\sqrt{s\beta_3}) + F(s)K_0(r\sqrt{s\beta_3}) + \frac{U_0}{s} \\ U_4(r,s) = E(s)I_0(r\sqrt{s\beta_4}) + F(s)K_0(r\sqrt{s\beta_4}) + \frac{U_0}{s} \\ U(r,s) = \int_0^\infty e^{-st}u(r,t)dt \end{cases} \tag{11}$$

where $A(s)$, $B(s)$, $C(s)$, $D(s)$, $E(s)$, $F(s)$, $G(s)$, and $H(s)$ are unknown coefficients; $I_0$ and $K_0$ are Bessel functions of the first and second classes of virtual variables of order zero, respectively; and $s$ is a complex parameter.

The undetermined coefficient in Equation (11) can be determined by considering the boundary conditions. Adhering to the principles of energy conservation and the second law of thermodynamics, the temperature and heat flux must remain continuous and equal at the interface of the two computed regions. Simultaneously, due to the relatively modest heat source intensity and the limited duration of the test, it is feasible to approximate the external surface of the concrete outer wall as having a constant temperature.

The heat source term directly influences the steel strand. The matrix equation for resolving the unspecified coefficients can be derived by applying the boundary conditions, as expressed in Equation (12).

$$AX = B \tag{12}$$

Here, the matrices $A$, $B$, and $X$ are defined according to Equation (13). To solve this matrix equation, the temperature distribution over time can be ascertained through multiple iterations employing numerical inversion techniques.

$$\begin{cases} A = \begin{bmatrix} I_1(0) & K_1(0) & 0 & 0 & 0 & 0 & 0 & 0 \\ \lambda_1 I_1(r_1\sqrt{s\beta_1}) & \lambda_1 K_1(r_1\sqrt{s\beta_1}) & -\lambda_2\sqrt{\frac{\beta_2}{\beta_1}} I_1(r_1\sqrt{s\beta_2}) & -\lambda_2\sqrt{\frac{\beta_2}{\beta_1}} K_1(r_1\sqrt{s\beta_2}) & 0 & 0 & 0 & 0 \\ I_0(r_1\sqrt{s\beta_1}) & K_0(r_1\sqrt{s\beta_1}) & -I_0(r_1\sqrt{s\beta_2}) & -K_0(r_1\sqrt{s\beta_2}) & 0 & 0 & 0 & 0 \\ 0 & 0 & \lambda_2 I_1(r_2\sqrt{s\beta_2}) & \lambda_2 K_1(r_2\sqrt{s\beta_2}) & -\lambda_3\sqrt{\frac{\beta_3}{\beta_2}} I_1(r_2\sqrt{s\beta_3}) & -\lambda_3\sqrt{\frac{\beta_3}{\beta_2}} K_1(r_2\sqrt{s\beta_3}) & 0 & 0 \\ 0 & 0 & I_0(r_2\sqrt{s\beta_2}) & K_0(r_2\sqrt{s\beta_2}) & -I_0(r_2\sqrt{s\beta_3}) & -K_0(r_2\sqrt{s\beta_3}) & 0 & 0 \\ 0 & 0 & 0 & 0 & \lambda_3 I_1(r_3\sqrt{s\beta_3} & \lambda_3 K_1(r_3\sqrt{s\beta_3} & -\lambda_4\sqrt{\frac{\beta_4}{\beta_3}} I_1(r_3\sqrt{s\beta_4}) & -\lambda_4\sqrt{\frac{\beta_4}{\beta_3}} K_1(r_3\sqrt{s\beta_4}) \\ 0 & 0 & 0 & 0 & I_0(r_3\sqrt{s\beta_3}) & K_0(r_3\sqrt{s\beta_3}) & -I_0(r_3\sqrt{s\beta_4}) & -K_0(r_3\sqrt{s\beta_4}) \\ 0 & 0 & 0 & 0 & 0 & 0 & I_1(r_3\sqrt{s\beta_4}) & K_1(r_3\sqrt{s\beta_4}) \end{bmatrix} \\ B = \begin{pmatrix} 0 & 0 & \frac{-I^2 R}{s\rho c V} & 0 & 0 & 0 & 0 & 0 \end{pmatrix}^T \\ X = \begin{pmatrix} A(s) & B(s) & C(s) & D(s) & E(s) & F(s) & G(s) & H(s) \end{pmatrix}^T \end{cases} \tag{13}$$

### *5.3. Analysis and Comparative Results*

By utilizing the derived theoretical solution, the temperature distribution of the controllable bonded electrothermal method can be computed iteratively through programming. Given that the prestressed strands can be considered as an infinite composite cylindrical wall within this calculation approach, the theoretical temperature value becomes independent of the length direction of the prestressed strands and solely dependent on the radius of the measuring point's section. This results in a unique solution once heating conditions, material properties, and material property states are established. The parameter inputs used for analyzing the theoretical solution are outlined in Table 4.

Of note, the prestressed reinforcement positioned at the center of the concrete beam is in closest proximity to the lower surface of the concrete beam, which leads to the most rapid heat dissipation and consequently, the lowest temperature. To ensure optimal performance of the controllable adhesive reinforcement, the temperature at this juncture should not fall below the designated design value. Illustrated in Figure 22 is a comparison curve showcasing the theoretical calculation results alongside experimental test results after heating at the midpoint.

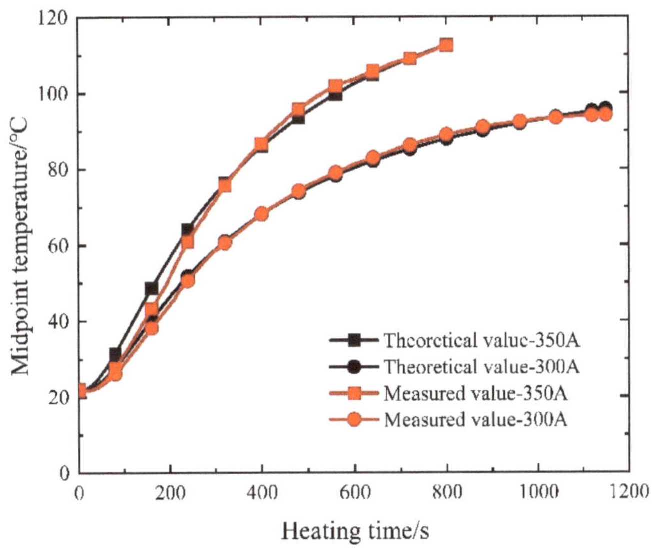


**Figure 22.** Comparison curve of the measured and theoretical temperature of the prestressed tendon.

Upon scrutinizing the curves presented in Figure 22, it becomes evident that the maximum relative errors between the theoretical and measured values at the midpoint, considering the two distinct current sizes, amount to 1.28% and 0.61%, correspondingly. Notably, as the heating duration extends, the error diminishes, implying a heightened degree of accuracy associated with the calculation method. However, it is worth noting that the measured temperature ultimately falls below the calculated temperature primarily due to the assumption of a constant wall temperature for the outermost concrete wall during computation, whereas a minor heat loss between the outermost concrete wall and the surrounding air transpires during the actual measurement.

Following the comparison between theoretical and experimental results, a parameter analysis is conducted to investigate varying currents, controllable bonded binder thicknesses, and steel strand diameters. This comprehensive analysis involves modifying the calculation program in accordance with numerical simulation's parameter variations.

5.3.1. Current Size

When adjusting the current density applied to the end face of the steel strand, it is essential to consider that the curing reaction of the controllable bonded binder is only initiated once the predetermined internal temperature is attained. Determining the necessary heating duration follows a constrained temperature rise approach. Commencing with an initial model temperature of 20 °C, the heating current systematically varies from 200 A to 500 A in 50 A increments. The resultant comparison between simulated and theoretical values, focusing on the time required for the midpoint of the prestressed reinforcement to reach 80 °C, is depicted in Figure 23.

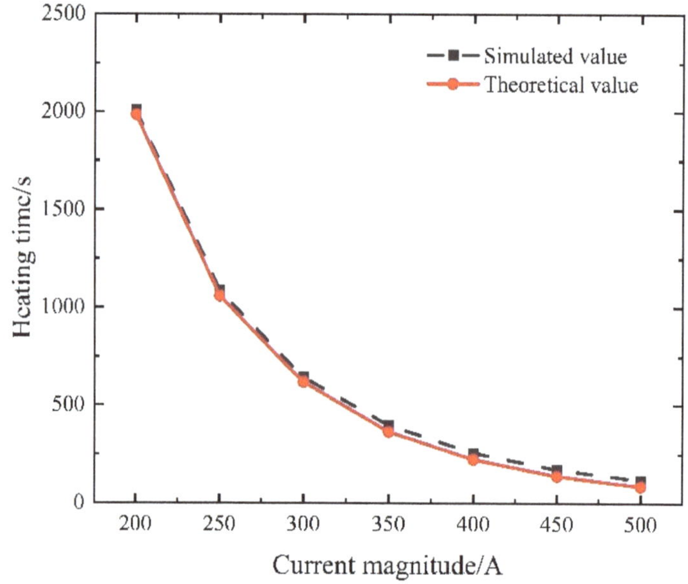


**Figure 23.** Comparison between simulation and theoretical results.

Examining Figure 23, it becomes apparent that a parallel trend is observed in both the numerical simulation and theoretical predictions as the current is varied. The maximum relative error, reaching 24.61%, materializes when the current is set at 500 A. This divergence is attributed to the assumption of a constant material thermal conductivity during calculations. In reality, material thermal conductivity increases with rising temperatures, hastening heat dissipation and necessitating extended heating times. When applying lower current values, the calculated results exhibit a closer alignment with the simulated outcomes. For instance, at 200 A, the relative error stands at a mere 1.09%.

5.3.2. Controllable Bonded Binder Thickness

The controllable bonded binder thickness was constrained within the range of 2 mm to 3.5 mm, employing an incremental step size of 0.25 mm. Under identical current and heating duration conditions, the temperatures corresponding to various controllable bonded binder thicknesses were computed. Subsequently, these results were charted as a scatter plot illustrating the temperature distribution, which was further subjected to a linear regression analysis.

The comparison depicted in Figure 24 reveals a congruence between the simulation and calculation outcomes, showcasing analogous trends. An augmentation in the controllable bonded binder thickness leads to a proportional increase in its thermal resistance, thereby causing a gradual elevation in the midpoint temperature of the prestressed steel strand.

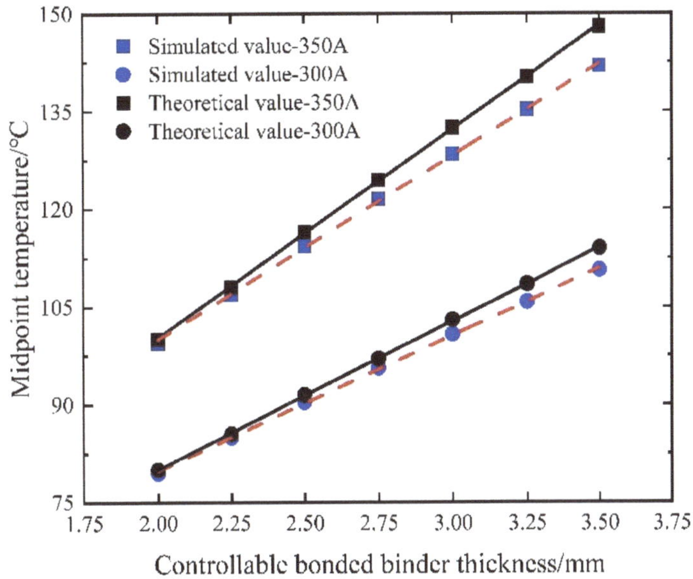


**Figure 24.** Midpoint temperature variation for different controllable bonded binder thicknesses.

The calculated temperatures tend to exceed the simulated values due to the constant thermal conductivity assumption made in the calculation. As a result, there is an increasing relative error between the two values with the controllable bonded binder thickness, reaching a maximum relative error of 4.23%. This observation highlights the effectiveness of augmenting the controllable bonded binder thickness for enhancing heating efficiency, especially under high current conditions.

5.3.3. Diameter of Steel Strand

In the role of a conducting medium, the diameter of the steel strand directly influences the magnitude of Joule heat while keeping material properties constant. Referring to previous calculations involving a 15.2 mm steel strand, an examination of commonly used steel strand diameters prompts an investigation into the temperature field induced by electric heating within prestressed strands with diameters of 17.8 mm and 21.8 mm when subjected to a current of 350 A during heating. The resulting comparison with numerical simulation results is visually presented in Figure 25.

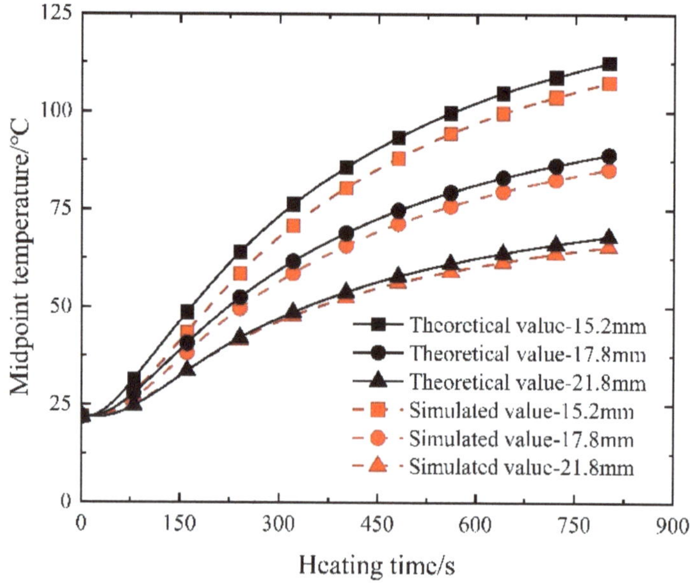


**Figure 25.** Temperature–time variation for electric heating of steel strands with different diameters.

As depicted in Figure 25, a concordant pattern emerges between the computed outcomes and the simulation results. Progressing from 15.2 mm to 17.8 mm and further to a 21.8 mm steel strand diameter, a discernible decrease in the electric heating rate transpires. Consistently, under equivalent heating conditions, the controllable bonded strands characterized by larger diameters achieve the heating stage at a swifter pace. It is noteworthy that the calculated results converge more closely with the simulated data as the diameter of the steel strand increases. This tendency is attributed to the gradual heating process and minimal alterations in material thermal parameters, aligning the calculation boundary conditions more closely with their simulated counterparts. Notably, the maximum relative error manifests at the 15.2 mm diameter, registering at 4.76%, thereby affirming the heightened reliability of the calculated outcomes.

## 6. Conclusions

The significance of this study lies in the fact that traditional retard-bonded prestressing requires several months of waiting for the binder to cure and reach its design strength after tensioning. However, by incorporating a latent-heat curing agent into the binder and activating it through electric thermal methods, rapid curing can be achieved in a matter of minutes, enhancing the strength. This makes it suitable for dynamic loading scenarios such as bridges and highways, and it effectively reduces the construction time.

Incorporating DC heating into controllable bonded prestressed reinforcement offers a feasible and controllable approach, with the critical caveat that meticulous regulation of the heating current is imperative to prevent potential adverse effects, ensuring the preservation of both structural integrity and material properties. By establishing a reasonable heating method, it takes only a few minutes to ensure the curing of the controllable binder. This is underscored by a maximum relative error of 1.73% in agreement with the proposed theoretical model. Moreover, optimizing heating parameters, specifically current size and binder thickness, demonstrates a significant reduction in the heating time, presenting a theoretical solution for parameter determination based on real material properties. However, it is essential to note that further research should address unconsidered factors such as contact resistance at heating terminals, the influence of anchor pads on heat generation, and thermal resistance between material layers. These enhancements highlight the expanding potential for electric heating applications in achieving controllable bonding and temperature-induced stress management within prestressed reinforcement for more efficient and reliable prestressed concrete structures.

(1) Feasibility and control of electric heating: The utilization of electric heating to regulate the temperature within controllable bonded prestressed reinforcement proves to be a viable and effective technique in terms of energy consumption, efficiency, and safety. However, precise control of the heating current is crucial to prevent the adverse effects caused by excessively high temperatures resulting from inadequate current, prolonged heating durations, or excessive current flow. This ensures the preservation of both the structural integrity and mechanical properties of the steel strand and concrete.
(2) Agreement between theoretical model and experimental data: The developed theoretical model for electric heating of controllable bonded prestressed reinforcement demonstrates remarkable agreement with the observed distribution of the temperature field. The maximum relative error of 1.73% highlights the accuracy achieved by this proposed theoretical framework.
(3) Optimization through current size and binder thickness: The analysis reveals that increasing the current size while judiciously augmenting the thickness of controllable bonded binder contributes to a reduction in heating time. Selecting an appropriate current size is pivotal in establishing a stable temperature environment conducive to the curing process.
(4) Theoretical solution for parameter determination: The theoretical solution method proposed in this study surpasses the experimental results, providing the capability to calculate the temperature field values based on actual material parameters in real-world applications. This facilitates the identification of optimal heating parameters.

Notably, the current study does not take into account contact resistance at heating terminals, the impact of anchor pads on heat generation, and thermal resistance between material layers. Addressing these factors in future simulations and tests would enhance our understanding of the intricacies of the electrothermal method. Meanwhile, with the continuous emergence of novel controllable bonded binders, the electrothermal method is poised to find an expanded utility in achieving controllable bonding. Furthermore, leveraging temperature-induced stress within prestressed reinforcement holds potential for enabling controlled tension, reducing prestress loss, and promoting more uniform dispersion of effective prestress.

**Author Contributions:** Conceptualization, N.J.; Methodology, X.X. and N.J.; Software, N.J.; Validation, X.X. and N.J.; Formal analysis, X.X. and N.J.; Investigation, X.X.; Data curation, N.J.; Writing—original draft, N.J.; Writing—review & editing, X.X.; Supervision, X.X.; Project administration, X.X.; Funding acquisition, X.X. All authors have read and agreed to the published version of the manuscript.

**Funding:** This research received no external funding.

**Data Availability Statement:** The data presented in this study are available on request from the corresponding author.

**Conflicts of Interest:** The authors declare no conflict of interest.

## References

1. Laura, A.; Antonio, B.; Giuseppe, D. Damage and collapse mode of existing post tensioned precast concrete bridge: The case of Petrulla viaduct. *Eng. Struct.* **2018**, *162*, 226–244.
2. Takashi, K.; Akira, Y.; Kumi, H.; Kazuhiro, F.; Takeshi, Y.; Norio, T. Immobilization of flame-retardant onto silica nanoparticle surface and properties of epoxy resin filled with the flame-retardant-immobilized silica. *React. Funct. Polym.* **2013**, *73*, 613–618.
3. *JG/T 370-2012*; Adhesive for Retard-Bonded Prestressing Steel Strand. Standardization Administration of the People's Republic of China: Beijing, China, 2012. (In Chinese)

4. *JGJ 369-2016;* Code for Design of Prestressed Concrete Structures. Standardization Administration of the People's Republic of China: Beijing, China, 2016. (In Chinese)
5. *JG/T 387-2012;* Filled Epoxy-Coated Prestressing Steel Strand. Standardization Administration of the People's Republic of China: Beijing, China, 2012. (In Chinese)
6. Xiong, X.Y.; Xiao, Q.S.; Li, X. Review of research on retard-bonded prestressed. *Build. Struct.* **2018**, *48*, 83–90. (In Chinese)
7. Yin, S.; Meng, F. Research on electric heating technology for prestressed reinforcement of old concrete Bridges. *Eng. Equip. Mater.* **2021**, *5*, 145–146. (In Chinese)
8. Santos, A.; Santos, P. Effect of surface preparation and bonding agent on the concrete-to-concrete interface strength. *Constr. Build. Mater.* **2012**, *37*, 102–110. [CrossRef]
9. Ahmed, H.; Aziz, O. Shear behavior of dry and epoxied joints in precast concrete segmental box girder bridges under direct shear loading. *Eng. Struct.* **2019**, *182*, 89–100. [CrossRef]
10. Diab, A.; Elmoaty, A.; Eldin, M. Slant shear bond strength between self compacting concrete and old concrete. *Constr. Build. Mater.* **2017**, *130*, 73–82. [CrossRef]
11. Valikhani, A.; Jahromi, A.; Mantawy, A.; Azizinamini, A. Experimental evaluation of concrete-to-UHPC bond strength with correlation to surface roughness for repair application. *Const. Build. Mater.* **2020**, *238*, 117753. [CrossRef]
12. Yeon, J.; Song, Y.; Kim, K.; Kang, J. Effects of Epoxy Adhesive Layer Thickness on Bond Strength of Joints in Concrete Structures. *Materials* **2019**, *12*, 2396. [CrossRef]
13. Çolak, A.; Çoşgun, T.; Bakırcı, A. Effects of environmental factors on the adhesion and durability characteristics of epoxy-bonded concrete prisms. *Constr. Build. Mater.* **2009**, *23*, 758–767. [CrossRef]
14. Czaderski, C.; Martinelli, E.; Michels, J.; Motavalli, M. Effect of curing conditions on strength development in an epoxy resin for structural strengthening. *Compos. Part B-Eng.* **2012**, *43*, 398–410. [CrossRef]
15. Moussa, O.; Vassilopoulos, A.; Castro, J.; Keller, T. Early-age tensile properties of structural epoxy adhesives subjected to low-temperature curing. *Int. J. Adhes.* **2012**, *35*, 9–16. [CrossRef]
16. Mays, G.; Hutchinson, A. *Adhesives in Civil Engineering;* Cambridge University Press: Cambridge, UK, 1992.
17. Sancaktar, E.; Jozavi, H.; Klein, R. The Effects of Cure Temperature and Time on the Bulk Tensile Properties of a Structural Adhesive. *J. Adhes.* **1983**, *15*, 241–264. [CrossRef]
18. Carbas, R.; Marques, E.; Silva, L.; Lopes, A. Effect of cure temperature on the glass transition temperature and mechanical properties of epoxy adhesives. *J. Adhes.* **2014**, *90*, 104–119. [CrossRef]
19. Islam, M.; Pickering, K.; Foreman, N. Curing kinetics and effects of fibre surface treatment and curing parameters on the interfacial and tensile properties of hemp/epoxy composites. *J. Adhes. Sci. Technol.* **2009**, *23*, 2085–2107. [CrossRef]
20. Cai, Y.; Gao, S.; Wang, F.; Zhang, Z.; Zhao, Z. Early hydration heat temperature field of precast concrete T-beam under steam curing: Experiment and simulation. *Case Stud. Constr. Mat.* **2023**, *13*, 8–16.
21. Jiang, Y.; Zhu, Y. Temperature field simulation of electric heating prestressed concrete component. *Concrete* **2011**, *2*, 10–16. (In Chinese)
22. Wei, J.; Wang, H.; Mao, J.; Zhu, Q.; Wang, F.; Xie, Y. Numerical simulation and test verification for temperature field of concrete continuous box girder bridges. *J. Southeast Univ. (Nat. Sci. Ed.)* **2021**, *51*, 378–383. (In Chinese)
23. Khan, M. Factors affecting the thermal properties of concrete and applicability of its prediction models. *Build. Environ.* **2002**, *37*, 607–614. [CrossRef]
24. Iman, A.; Payam, S.; Zahiruddin, F.; Norhayati, B. Thermal conductivity of concrete—A review. *J. Build. Eng.* **2018**, *201*, 81–93.
25. Wang, W.; Liu, B.; Venkatesh, K. Effect of Temperature on Strength and Elastic Modulus of High-Strength Steel. *J. Mater. Civ. Eng.* **2012**, *25*, 174–182. [CrossRef]
26. Venkatesh, K.; Sonali, K.; Wasim, K. Effect of Temperature on Thermal and Mechanical Properties of Steel Bolts. *J. Mater. Civ. Eng.* **2011**, *24*, 765–774.
27. Zhou, H.; Li, G.; Jiang, S. Experimental Studies on the Properties of Steel Strand at Elevated Temperatures. *J. Sichuan Univ.* **2008**, *40*, 106–110. (In Chinese)
28. Shakya, A.; Kodur, V. Effect of temperature on the mechanical properties of low relaxation seven-wire prestressed strand. *Constr. Build. Mater.* **2016**, *124*, 74–84. [CrossRef]
29. Hou, X.; Zheng, W.; Kodur, V.; Sun, H. Effect of temperature on mechanical properties of prestressing strands. *Constr. Build. Mater.* **2014**, *61*, 24–32. [CrossRef]
30. Kim, K.; Jeon, S.; Kim, J.; Yang, S. An experimental study on thermal conductivity of concrete. *Cement Concrete Res.* **2003**, *33*, 363–371. [CrossRef]
31. Churchill, S.; Chu, H. Correlating equations for laminar and turbulent free convection from a vertical plate. *Int. J. Heat Mass Tran.* **1975**, *18*, 1323–1329. [CrossRef]
32. Stehfest, H. Algorithm 368 numerical inversion of Laplace transforms. *Commun. ACM* **1970**, *13*, 47–49. [CrossRef]
33. Wang, P.; Lin, S.; Tu, S. A survey of fractional-calculus approaches to the solutions of the Bessel differential equation of general order. *Appl. Math. Comput.* **2006**, *187*, 544–555. [CrossRef]

34. Lin, S.; Ling, W. A Simple Fractional-Calculus Approach to the Solutions of the Bessel Differential Equation of General Order and Some of Its Applications. *Comput. Math. Appl.* **2005**, *49*, 1487–1498. [CrossRef]
35. Özisik, M.N. *Heat Conduction*, 2nd ed.; John Wiley & Sons: New York, NY, USA, 1993.

MDPI AG
Grosspeteranlage 5
4052 Basel
Switzerland
Tel.: +41 61 683 77 34

*Materials* Editorial Office
E-mail: materials@mdpi.com
www.mdpi.com/journal/materials

www.ingramcontent.com/pod-product-compliance
Lightning Source LLC
Chambersburg PA
CBHW041347210926
51850CB00013B/3013
*9783725839162*